Brazil is a country of immense diversity. Its continental dimensions contain the most important industrial complex of the south as well as the largest rainforest reserve in the world, the Amazon. The 1970s have witnessed a period of immense economic growth in Brazil, yet more than half the population live in poverty. In this major new textbook, Professors Bertha Becker and Claudio Egler examine these contemporary dilemmas by exploring the process of Brazil's entry into the capitalist world-economy. They trace this development from the country's origins as a Portuguese colony to its current status as a regional power in Latin America and the eighth-largest world economy. Becker and Egler also explore the different phases of Brazil's political economy, the role of the state in a newly industrialized country, the legacy of conservative modernism and the dimension of Brazil's current debt crisis. In their conclusion, they assess how Brazil's challenges, as a semiperipheral country and a regional power, may affect the restructuring of the world capitalist system.

Professors Becker and Egler combine geography, history, economics, and political science in a comprehensive view of Brazil's development and this innovative and compelling approach enables comparative analysis with other countries. *Brazil: a new regional power in the world-economy* will be widely read by students and specialists in geography, Latin American history, political science, development economics, urban and regional planning, and public administration. It will also be an invaluable reference source for journalists, government analysts, and policymakers in international development agencies.

Geography of the World-Economy

Series Editors:

PETER TAYLOR   University of Newcastle upon Tyne (General Editor)
JOHN AGNEW   Syracuse University
CHRIS DIXON   City of London Polytechnic
DEREK GREGORY   University of British Columbia
ROGER LEE   Queen Mary College, London

A geography without knowledge of place is hardly a geography at all. And yet traditional regional geography, underpinned by discredited theories of environmental determinism, is in decline. This new series *Geography of the World-Economy* will reintegrate regional geography with modern theory and practice – by treating regions as dynamic components of an unfolding world-economy.

Geography of the World-Economy will be a textbook series. Individual titles will approach regions from a radical political-economic perspective. Regions have been created by individuals working through institutions as different parts of the world have been incorporated in the world-economy. The new geographies in this series will examine the ever-changing dialectic between local interests and conflict and the wider mechanisms, economic and social, which shape the world system. They will attempt to capture a world of interlocking places, a mosaic of regions continually being made and remade.

The readership for this important new series will be wide. The radical new geographies it provides will prove essential reading for second-year or junior/senior students on courses in regional geography, and area and development studies. They will provide valuable case-studies to complement theory teaching.

Brazil: a new regional power in the world-economy

Brazil: a new regional power in the world-economy

Bertha K. Becker
Professor, Department of Geography,
Federal University of Rio de Janeiro

Claudio A. G. Egler
Professor, Department of Geography,
Federal University of Rio de Janeiro

CAMBRIDGE
UNIVERSITY PRESS

CAMBRIDGE UNIVERSITY PRESS
Cambridge, New York, Melbourne, Madrid, Cape Town, Singapore, São Paulo, Delhi

Cambridge University Press
The Edinburgh Building, Cambridge CB2 8RU, UK

Published in the United States of America by Cambridge University Press, New York

www.cambridge.org
Information on this title: www.cambridge.org/9780521379052

First published 1992
Re-issued in this digitally printed version 2009

A catalogue record for this publication is available from the British Library

Library of Congress Cataloguing in Publication data

Becker, Bertha K.
Brazil, a new regional power in the world-economy / Bertha K.
Becker, Claudio A. G. Egler
    p.   cm. – (Geography of the world-economy)
Includes bibliographical references.
ISBN 0 521 37008 6 (hardback). – ISBN 0 521 37905 9 (paperback)
1. Brazil – Economic conditions. 2. Brazil – Foreign economic
relations. I. Egler, Claudio A. G. II. Title. III. Series.
HC187.B3897 1992
337.81 – dc20   91–27035   CIP

ISBN 978-0-521-37008-0 hardback
ISBN 978-0-521-37905-2 paperback

Contents

Figures

Tables

Preface

When recently asked about the most remarkable change he had seen since his first visit to Brazil in 1958, John Kenneth Galbraith responded: Brazil's emergence, in spite of all the difficulties, as one of the world's great powers (*Veja*, June 20, 1990). This declaration might seem strange about a country which is mostly known in the international media for Carnival and football, and fills today's headlines for having one of the world's largest foreign debts and the most unequal income distribution on the planet, not to mention the ecological polemic over the destruction of the Amazon forest.

Actually, Brazil is an emerging power with all the attendant difficulties, which include the poverty of more than half of its population, aggravated by a profound economic crisis. It is difficult to know Brazil's reality and its position in the world because of the accelerated rhythm of changes in a country whose continental dimensions generate multiple times and spaces, official statistics which are obsolete on the day of their publication, and the restriction of information during long years of authoritarian rule. This difficulty is apparent in polarized attitudes and disparate prognoses for Brazil's future.

This is a book of regional geography, and proposes to view Brazil as an integrated and differentiated part of the larger whole: the capitalist world-economy. The concept of region is associated with the work of the geographer. Ignoring it would be to lose a sign which identifies geography among the other sciences. Rethinking the concept of region today means contributing to overcoming the crisis in the social sciences and to collaborating, as geographers, in understanding the contradictions and impasses of the contemporary world. Nevertheless, critical reform of geography identified the concept of region with the empiricist tradition of the classical school, attaching to it the stigma of ideographic methods, and ending up by throwing the baby out with the bath water.

This book's task is to relate space and time; that is, to write a regional geography, in the world-economy perspective, through examining the process of Brazil's insertion into the world capitalist system – which simultaneously means its individualization as a region. It is an attempt to participate in constructing a new path for regional geography, as well as to develop a world-economy perspective which is more sensitive to places.

Various people have collaborated to make this book possible, and we thank them. Peter Taylor initiated and stimulated this undertaking. William Savedoff had the arduous task of translating the text into English and offered many useful comments. Maria Helena Lacorte and Rodolfo Bertoncello also gave helpful criticism. The National Council for Scientific and Technological Development (CNPq) and the Financer of Studies and Projects (FINEP) have been sponsoring our research on Brazil for years. Last but not least, our companions and children gave the essential tolerance and affection to support the long months of preparing the text, and we dedicate this book to them.

Rio de Janeiro, June 1990

1
The ambivalence of an emerging power

The cadets of the British Royal Air Force are being trained on the EMB-312 Tucano, a turboprop aeroplane developed by the Brazilian aeronautics company EMBRAER and assembled under a contract with Great Britain involving transfer of technology. This illustrates the success of a nascent arms industry which places Brazil among the ten largest producers of weapons in the world, an unprecedented situation for any other Latin American country. In terms of war, the country has made itself fully modern, but in terms of peace it has barely reached the threshold. Average life expectancy in Brazil is sixty-five years, inferior to that in the neighboring countries of Paraguay (sixty-seven years), Argentina (seventy years), and Uruguay (seventy-one years).

Brazil is little known, even by those who live and work there. The speed with which transformations took place in the last thirty years, combined with systematic control of information by the authoritarian regime, have made it difficult to comprehend its real dimensions. Brazil is not a great power, which the military so ardently desire and consider to be the country's manifest destiny. Nor is it a member of the "Third World," as preached by its most simplistic critics. Rather, Brazil is an example of an emerging power within a regional context, marked by contradictory features.

The ambivalence of a regional power has to do with three levels of power. The first involves growing in a space submitted to the hegemony of a world center, in this case, Latin America under the dominant influence of the United States. The second reflects shifting pretensions by competitors in the regional context, in this case, Argentina, which was until recently the most important economy in South America. The third level expresses the political control over territory and society which, within Brazil, assumed the form of a national authoritarian project.

This chapter portrays Brazil's contradictory features within the world context. It presents the concepts of historic capitalism and the world-system, developed originally by Immanuel Wallerstein, which make it possible to situate Brazil in the semiperipheral sector of the capitalist world-system. Finally, it utilizes some elements in Latin American and Brazilian social thought in order to characterize the specificities of Brazil as a region within the world-economy.

An unknown continent

Brazil is a country of multiple times and multiple spaces. The speed with which technological innovations are incorporated is extremely rapid in certain parts of its territory, taking place alongside primitive living conditions whose rhythms are determined by nature in immense extensions. Grand national television networks, similar to the North American pattern, daily establish the bridge between past and future, between isolated *garimpeiros* (prospectors) in the jungle searching for El Dorado and managers of large multinational corporations seated on Avenida Paulista, the "Wall Street" of Brazil.

Accelerated timing

As part of the world-economy, Brazil is one of the most dynamic segments when economic indicators are considered. Its historic rates of gross national product (GDP) growth are comparable with the advanced economies since the end of the last century. Since 1940, GDP growth averaged 7 percent per year, attaining 11 percent annually between 1967 and 1973 – the years of the "economic miracle" which coincided with evident signs of slackening growth in the rest of the world (Table 1.1).

Accelerated urbanization is the most obvious expression of this process in terms of space. In contrast to the majority of Latin American economies, Brazil does not have an extreme concentration of activity in a single grand national metropolis. Instead, Brazil has a constellation of nine metropolitan areas with more than one million inhabitants each and a large number of medium and small cities dispersed throughout its territory.

Its recent industrialization distinguishes Brazil from the rest of Latin America; in this regard it far surpassed Argentina and was accompanied with less intensity by Mexico. The association with

Table 1.1. *Average rate of GDP growth in constant prices (%)*

Countries	1870–1913	1913–50	1950–73	1973–83
United States	4.2	2.8	3.7	1.9
Germany	2.8	1.3	5.9	1.6
Japan	2.5	2.2	9.4	3.7
Mexico	2.0	2.7	6.6	4.6
Brazil	2.3	4.9	7.5	4.5

Source: Adapted from Maddison 1982, 1985.

international capital was a common feature in the region's development but, in Brazil, the state had a decisive role in accelerating the pace of growth, advancing the private sector, and maintaining high rates of investment. On the other hand, Brazil is also one of the largest debtors within the world financial system in absolute terms. Its total debt was approximately US$112 billion in 1988, a balance equivalent to a little more than one-third of its GDP.

The Latin American industrialization model, based on import-substitution, sought to manage the domestic market as the principal attraction for large transnational corporations. Brazil reached more advanced stages of this process, and successfully consolidated a diversified industrial base – due largely to the potential of its economy, whose capacity to attract capital was made viable and amplified by the state.

A moving frontier

Brazil is a continent. This is a fundamental difference in comparison with its Latin American neighbors. Its territorial extent places it fifth among the nations of the planet in terms of size, with an area of 8.5 million square kilometers and a population of 155 million inhabitants in 1990. Its demographic density of 18.2 inhabitants per square kilometer leaves immense potential for growth, since it is more sparsely settled than the United States and comparable to the Soviet Union.

Brazil's resource potential is amplified by the availability of usable space, resulting from its geographical position. Brazil, which includes two-thirds of the South American continent, is certainly the largest country situated in the intertropical band. The grand reserve of land in Brazil is the greatest rain forest on the planet, the Amazon, with its immense variety of species. Nevertheless, the delicate ecological

Table 1.2. *Income distribution – selected countries*

Countries	Years	Bottom 20%	Lower 40%	Upper 20%	Top 10%
India	1984–85	9.8	22.8	39.4	25.4
Peru	1985	4.4	12.9	51.9	35.8
Brazil	1983	2.4	8.1	62.6	46.2
Venezuela	1987	4.7	13.9	50.6	34.2
Italy	1986	6.8	18.8	41.0	25.3

Source: World Bank 1990.

balance of the Amazon challenges Brazilian society and world science to find appropriate forms of occupation.

Natural conditions are important, but they are not determinant. While Argentina maintained its agro-pastoral orientation to complement the world center, Brazil, which historically played a comparable complementary role, has today diversified its export pattern of agricultural "commodities." Originally a large coffee producer, Brazil is currently the second largest exporter of soy and derivatives, with the advantage of placing its product on the market between North American harvests. Although little known in Brazil only fifteen years ago, soy has conquered the ecological barrier of the *cerrados* (central savannahs) and spread rapidly, thanks to investments in genetic improvements and the development of cultivation treatments. In 1975, the *cerrados* were responsible for producing almost 6 percent of Brazilian soy; in 1982 they produced 22 percent of the national crop; and with the large harvest of some 8 million tons of soy in 1987–88, they accounted for 44.5 percent of the national total.

The Brazilian economy grew, and continues to grow, through an impressive capacity to incorporate new land rapidly. The total area occupied by agricultural establishments was 198 million hectares in 1940, jumping to 365 million hectares in 1980 – which already represents about half of the net area available for cultivation and ranching.

The role of the frontier and access to land are fundamental in distinguishing Brazil from its Latin American peers, configuring the agrarian question in a peculiar way. In Mexico, for example, the scarcity of arable land has been a problem since the beginning of industrialization. This made it necessary to confront the rigidity of agrarian structures and mobilize state resources to increase the production of food and agricultural raw materials. The agrarian

Table 1.3. *Ethnic situation*

Ethnic groups	Population (%)	Illiteracy (%)	Median income (US$/month)
Whites	56.6	12.3	214.00
Mulattoes	37.2	29.0	100.00
Blacks	5.6	29.5	87.00
Asians, and without declaration	0.6	7.4	377.00

Source: IBGE 1990b.

reform, then, was a requirement of the industrialization process itself. In Brazil, by contrast, the supply of agricultural products was guaranteed by the incorporation of new lands, without touching the pre-established structure of landownership which constituted the base of power for the dominant groups.

The large rural properties in Brazil, inherited from the slave latifundios, were a basic instrument for maintaining the reproduction of the labor force in conditions close to subsistence, lowering the general wage level in the economy. The caloric consumption per inhabitant in Brazil is 2,657 per day, lower than the values for Iraq (2,891), Iran (3,115), or Turkey (3,218). The hourly wage is among the lowest in the world – about US$1/hour – while in countries like the United Kingdom and France the average hourly wage is around US$17.

Brazil is a rich country of poor people. The brutal social discrimination in appropriating the benefits of economic dynamism is a dominant feature of Brazilian society, even when compared with the rest of Latin America. It is one of the few economies in the world where the richest 10 percent control almost half the national income, and any indicator of social welfare reflects this situation (Table 1.2).

This discrimination suffuses Brazil's social structure from top to bottom. Sexism is expressed by the fact that 67.1 percent of Brazilian women over ten years of age have no income, while only 24.7 percent of the men face this situation. Blacks and mulattoes, who represented some 45 percent of the Brazilian population in 1987, face social and economic discrimination in terms of social mobility. They make up a grand contingent of labor power with low professional qualifications, contrasting with the Asian immigrants and their descendants, mainly Japanese (Table 1.3). Ethnic discrimination is also present in relation to the remaining 200,000 indigenous people who survived the colonizing massacres; their rights are restricted and their capacity for self-determination is subordinated to bureaucratic tutelage.

In sum, Brazil carries the confrontation between North and South within a national economy. The contradictions of historic capitalism in Brazil assume a paradigmatic character, and the current crisis only accentuates its ambivalence, exposing the vulnerability of its power and its fragility as a nation. To understand Brazil's significance, it is necessary to define its position in the capitalist world-economy.

Historic capitalism, world-economy, and semiperiphery

For Wallerstein (1983), capitalism is above all a historic social system. To understand its origins, functioning, and current perspectives, we must look at the present reality. This premise is fundamental to specify the historicity of the geographic fact and its interpretation. This is not an attempt to find universal and atemporal explanations but, to the contrary, an effort to define and reaffirm the historic character of the social processes, and also the geographic forms which they create. The Brazil which will be described here is a region historically inscribed in the world-economy.

Rescuing history

The concept of historic capitalism is based upon the conceptions of Braudel (1979) concerning the *longue durée* of historic time, breaking the limits imposed by division into hermetic and isolated periods. Historic capitalism has its own dynamic which, from the start, is counterposed to a positivist vision of social progress. This is not "historicism," i.e. a deterministic vision of history, but rather historicity. There is nothing peculiar in the capitalist way of producing merchandise which would identify it as "social progress" in relation to earlier modes of production, much less that it exhausts itself.

The dynamic of historic capitalism occurs through long-lasting cycles of growth and recession. There is considerable evidence today that, at least since the end of the eighteenth century, the world-economy has passed through four great cycles. These have phases of growth (A) and stagnation (B) and were called "cycles" or "long waves" by Kondratieff (1935), the Russian historian who first identified them. The approximate dates of these cycles are:

	A	B	
I	1780–90	1810–17	1844–51
II	1844–51	1870–75	1890–96
III	1890–96	1914–20	1940–45
IV	1940–45	1967–73	?

A well-accepted explanation, although by no means definitive, focuses on the contradiction between the immediate interests of firms on the one hand, and the general requirements of business on the other. In phases of growth, profits are greater and firms tend to over-invest, generating crises of overproduction.

The correlation with technological changes is high. Growth in the A phases of successive cycles corresponds, respectively, to the Industrial Revolution (I), the expansion of the railroads and steel (II), chemical and electrical innovation (III), and the aerospace and electronic developments of the current cycle (IV).

There are also important political and spatial correlations. According to Wallerstein (1979), the four Kondratieff cycles when placed in political context can be described as "two-paired Kondratieffs." The first pair covers the nineteenth century, involving the rise and decline of English dominance. The second describes a similar trend for the United States in the current century – not that these represent equivalent forms of domination.

As for the period before 1780, there is little evidence for such cycles. Nonetheless, it is possible to identify "logistic waves," of about 300 years. Wallerstein recognizes two of these:

A		B
c. 1050	c. 1250	c. 1450
c. 1450	c. 1600	c. 1750

The first of these corresponds to the rise and fall of feudal Europe, and the second covers the emergence and crisis of European agricultural and mercantile capitalism, from which the modern world-economy emerged.

The conception of a capitalist world-economy as a universe of analysis which is historically and geographically determined breaks with the spatial imprecision of positivist origins, such as earth surface or landscape. The point of departure for constructing this model is the conception of a world-economy, consisting essentially of a single world capitalist market. This means not only that merchandise is produced primarily for the market rather than for use, but also that the development of exchange with the exterior and the existence of a world market are inherent conditions for the emergence and development of the capitalist mode of production. Wallerstein is emphatic in affirming that "capitalism was from the beginning an affair of the world-economy and not of nation-states" (1979: 19). The dynamic vector of the world-economy lies directly in the formation and development of this world market, both the origin and product of capitalism itself.

Meanwhile, if there is only one world market, it is based on a multiplicity of national states. "As a formal structure, a world-economy is defined as a single division of labour within which are located multiple cultures" (Wallerstein 1979: 159). This is a second element of Wallerstein's perspective, for were this not the case there would be only a single world-empire, that is, an immense territory subordinated to a single political entity.

The role of the nation-state is to distort the functioning of the world market in the interest of classes or social groups. The stronger the state machinery, the greater its capacity to influence the market's operation on a significant geographic scale, and even beyond its territorial limits. In this way, the nation-state is a political and economic instrument utilized by dominant classes and sectors, regional or national, to maintain parts of the world-economy under their control. They seek to control not only the markets for goods and products, but also, and principally, the labor-force market.

The first process of industrialization in Great Britain was exceptional because its economic advantages relative to eventual competitors permitted a relatively small role for the state. In all other cases, such as the United States, Germany, and Japan – referred to as "late industrialization" – the process counted on active state policy and territorial support for "national" enterprises.

The result of this active state participation in the struggle between monopolies could only be discharged in open conflicts, such as occurred in the two world wars from which North American hegemony over the world-economy emerged. The globalization of the dollar as an international currency and of the North American patterns of production and management expressed in Fordism and Taylorism unleashed a double process in the world-economy. On the one hand, multinational corporations whose interests define planetary space of operation grew and expanded rapidly; on the other hand, nation-states proliferated, following the process of decolonialization which redivided territories into ever smaller parcels, ever more distant from the "classical" concept of a nation (Hobsbawm 1977).

The territorial dimension of the world-economy

This complex network of relations formed a spatial structure which, in Wallerstein's view, is not exhausted by the classic model of center/periphery because the capitalist world-economy requires a semiperipheral sector. The semiperiphery assumes a fundamental role in the functioning of the world-economy, not so much economic as political, helping to stabilize the world-system.

The existence of the third category means precisely that the upper stratum is not faced with the unified opposition of all the others because the middle stratum is both exploited and exploiter. It follows that the specific economic role is not all that important, and has thus changed through the various historical stages of the modern world-system. (Wallerstein 1979: 21–23)

The concept of semiperiphery has an important aspect: the role of the state in politicizing the economy. "The direct and immediate interest of the state as a political machinery in the control of the market (internal and international) is greater than in either the core or the peripheral states, since the semiperipheral states can never depend on the market to maximize, in the short run, their profit margin" (Wallerstein 1979: 72).

The role filled by public funds in a semiperipheral economy is a peculiar one, since the state is a fundamental instrument for the accumulation process and is directly present in levering growth as well as in productive activities. But the other side of the coin is also important: by adopting the role of grand financier, the state ends up assuming the liabilities of the economy, becoming the debtor *par excellence*, fulfilling a role which historically was left to the investment banks. In the downturn of the cycle, the ambivalent character of the semiperiphery is apparent, since the economic crisis takes on the character of a profound political crisis.

The semiperiphery is the synthesis of the contradictions of historical capitalism within the same national economy. It is the *locus* of the profound structural heterogeneity accumulated by capitalism through its long history, of which Brazil is a magnificent example. But the category of semiperiphery does not exhaust the specificity of Brazil as a regional power. It is necessary to historicize it.

An authoritarian path toward modernity

Latin America is the oldest periphery of the world-economy. It is part of the process which formed and developed the capitalist world-system, oriented toward the production of high-value merchandise for Europe ever since the beginning of colonization. Divided between Spain and Portugal, Latin America's economic formation was marked by mercantilism and its society was molded in the image of Iberia. Its underlying development was intimately associated with the dynamic of the centers of accumulation in the world-economy – first, Great Britain and later the United States. Latin American countries participated in the resulting international division of labor as economies exporting primary materials.

Nevertheless, the general movement of the world-economy should not obscure the importance of internal factors in Latin America's formation. This formation is the product of a complex interplay in which the international context and the domestic situation both played significant parts. Their comparative weight, however, varied over time. In this sense, the task of reconsidering Latin American history has advanced in recent years. Its history cannot be described as a backward stage of development, nor as a mere extension of imperialist metropolises, but rather as an integral part of capitalism, inseparable and specific.

A certain capitalism

The *Economic Survey of Latin America – 1949* by the Economic Commission for Latin America (ECLA, also known by its Portuguese acronym CEPAL), marked the birth of the ECLA perspective of political economy. It was best represented in Brazil by Celso Furtado, and it clearly delineated the specific problematic of peripheral capitalism. The *Survey*, coordinated by Raul Prebisch, developed from evidence that Latin American growth relied directly upon the export sector, which furnished the foreign exchange necessary to import manufactures. This center–periphery structure tended to perpetuate itself, in that its dynamic was controlled by decisions made in the center and was accentuated by the deterioration of the terms of trade. In this context, only national industrialization would be capable of breaking the vicious cycle of underdevelopment.

The inability of industrialization to achieve national redemption through import-substitution encouraged new theories which sought to understand the roots of underdevelopment in the relations of dependence between center and periphery. One of these visions is Gunder Frank's concept of "the development of underdevelopment" (1967) which, treating the question on a general level, solved neither the Latin American problem nor the Brazilian one, as he himself recognized some years later. On the other hand, the formulation by F. H. Cardoso and E. Falletto (1970) contributed to a critical deepening of the concept of dependence. This formulation considers the issue in terms of the capitalist mode of production's formation and development in Latin America. It introduces the idea that the social dynamic in Latin America is determined in the first instance by internal factors and in the last instance by external factors, from the moment at which a national state is established.

Although accepting the statements of Cardoso and Falletto, a group of Brazilian economists criticized their periodization, which is restricted by the ECLA hypotheses. To overcome this limitation, it is necessary to think of Latin American history as the formation and development of a certain kind of capitalism, a capitalism which was born late (Mello 1982). Its origin lies in the old colonial system; it grew as a national mercantile economy; and the later generalization of wage labor in the agrarian-export sector was not matched by development of the specific productive forces of capitalism, which were only consolidated later with industrialization. This, in turn, had two basic constraints. First, Latin America depended directly upon the export sector for its accumulation; second, Latin America entered the world market at a time when it was already dominated by large firms.

The history of late capitalism, whose most elaborate expression is Brazil, is not a unique case nor a mechanical realization of pre-fixed stages. It is the history of capitalism in Latin America which, conversely, is also the history of the world-economy itself, as one of its parts and forms of sustenance.

A certain authoritarianism

The effort to understand populism as a form of exercising state power seems to be a landmark in the analysis of the authoritarian character of late capitalism in Latin America. Industrialization and the emergence of an urban proletariat put the inherited political structures of the agrarian-export past in check. They attributed a new role to the state in regulating relations between capital and labor. The state engaged wide social segments in its proposal for national development through a direct relation between charismatic leaders and the "people," without intermediation by political structures of representation (O'Donnell 1973; Germani 1977; Weffort 1978).

The crisis in the populist alternative unveiled the coercive essence of the political system and the nature of the Latin American nation-state. Innumerable studies show that its attributes are not limited to the delimitation of a domestic market for the nascent bourgeoisie but advance in the direction of imposing upon society its image of the nation. In this process, a fundamental component of Latin America's history stands out: that of state-building ahead of nation-building.

But authoritarianism has old roots in Latin America. While in the majority of European countries the organization of the state came after the existence of a society which was more or less organized, in Latin America the opposite occurred. The presence of the military–religious–bureaucratic apparatus guaranteed the monopoly over commerce; controlled the concession of lands; and legitimated slave and servile labor as local power bases for the grand "lords" in the colonies. When oligarchies and domestic mercantile interests conquered this apparatus, the newly constituted independent states did not seek to define national markets. Rather, they tried to make articulation with Great Britain viable, without breaking the hierarchical and authoritarian social structure which maintained them.

In Brazil, this situation assumed peculiar contours. The rest of Latin America was fragmented by disputes between *caudilhos*. In Brazil, political unity was maintained through a slavocratic pact which overcame the divergences between diverse local oligarchies, and defined an original role for the empire: that of guaranteeing the reproduction of the slave system in spite of British pressures to eliminate it.

The maintenance of a precarious equilibrium between the forces of the world market and the interests of dominant national groups conferred increasing tasks upon the state, which assumed a decisive role at the moment of industrialization. Throughout this process, a powerful state bureaucracy grew and consolidated itself, exercising political domination over civil society which was little organized and destitute of channels of representation. In this bureaucracy, the armed forces had a significant role, although with different levels of professionalization. Conservative modernization is the Latin American path to modernity, where the state negotiates with private groups for the maintenance of their privileges and their inclusion or exclusion from public matters, in exchange for their supporting the modernization project "from above."

Brazil is without equal in exemplifying this path, impelled by the military's modernization project. Brazil's flag carries the positivist slogan "Order and Progress," which affirms the defense of a hierarchical organization of society and envisions a future construction of the nation. In Brazil, nation-building as a historically constructed political ideology of the modern nation-state, involving an ideal image of how society should be organized, shows the ideological predominance of the collective individual over the collection of individuals and favors authority over solidarity (Reis 1983).

The other side of the coin is revealed in the distance between official institutions and those in society who seek their own paths, many times extra-legally. Eclecticism surpasses consistent ideology, pragmatism orients political decision making with weighty historical precedents, and results in powerful constraints upon full authoritarianism as much as upon democracy (Lamounier 1988). Hence, the ambivalence of the semiperiphery is also apparent on the political and ideological plane.

A certain territoriality

The discourse of national integration which assumed an elaborate form at the beginning of the 1970s helps to unveil the role of territoriality in the construction of Brazilian authoritarianism. The incorporation of "empty spaces" into the national domain was an essential part of the geopolitical project of modernization and of the country's rise as a regional power. It revealed a new meaning of territory in the mediation between state and society. It became a symbolic resource for constituting the collective individual in detriment to a national community of citizens.

Integration as a territorial ideology substituted the discourse of national unity which historically served to strengthen state-building. The label of separatism, as a way of suffocating political opposition movements, was utilized widely in Brazil from the beginnings of industrialization. Actually, the politics of territory preceded the territory of politics in the origin of the Brazilian state, which inherited the responsibility of maintaining the extensive limits established by the colonial Portuguese.

With independence, in contrast to Spanish Latin America which fragmented into a multiplicity of republics, the Brazilian monarchy not only maintained political unity but also guaranteed the territorial integrity of the old colony. The Brazilian empire inherited and reproduced the features which constituted the patrimonial Portuguese state, where the King, being situated above all his subjects, was the "lord" over the largest patrimony of land in the kingdom, and governed with a corps of functionaries recruited according to its exclusive choice and charged with maintaining and expanding the domains. Within these domains, the distinction between public and private goods was poorly maintained (Faoro 1959).

The control of strategic positions as an instrument for dominating territory is at the root of the formation of Brazilian geopolitics. This geopolitics always had a military logic and was present in the strategy for conquest and defense of the colony's territory by the Portuguese Crown during three centuries. It found continuity during the empire in the disputes for control over the grand basins of the Prata and the Amazon Rivers, perfecting itself in the definition of territorial limits at the turn of the century and, finally, consolidating itself in the construction of Brasília as a logistical base for definitive occupation of the backlands.

In contrast to Turner's view of the grand frontier as a key element in the construction of US democracy, in Brazil the frontier was associated with authoritarianism (Velho 1979). The availability of land favored pacts between oligarchies, allowing the creation of new latifundios without threatening the older territories and incorporating new areas into the agro-mercantile domain (Becker 1982). The conquest of the backlands was a symbol of extending "Order and Progress" to the "unproductive and lawless lands," justifying violent and predatory forms of occupation.

On the other hand, the backlands are the stage for historic struggles of resistance by indigenous settlements, *quilombos* – places of refugee slaves, and the territories of the poor. They were decimated. Nonetheless, the awareness that territoriality is a fundamental part of conquering citizenship and constructing the nation continues to permeate the social and political movements of Brazil today.

In sum, territorial and state-building rather than nation-building marks Brazil's authoritarian path toward modernity. Brazil not only carries within itself the conflict between North and South; it also has something more, an incognito represented by the Amazon, one of the planet's last frontiers.

This book has a task and an ambition. Its task is to describe and to analyze the process of Brazil's insertion into the capitalist world-economy, from its origins as a Portuguese colony to the attainment of its current status as a regional power. It involves a twofold process, on the one hand showing the effects of the dynamic of the world capitalist system upon its socio-spatial formation, and on the other hand revealing the local factors which influence this formation and the design of its regions. The book's ambition is to contribute theoretically to the elucidation of the role of the semiperiphery, where political factors are decisive, in the stability and trajectory of the triadic structure of the world-economy.

Brazil's historical evolution, the construction of its territory, and the building of the state ahead of the nation, are treated first (Chapter 2). In the next chapter, the process by which Brazil's regions were constituted is detailed historically as a result of the selective expansion of the world-economy (Chapter 3). Brazil's ascendance as a regional power is analyzed as a product of the recent transformations of the capitalist system combined with a national geopolitical project (Chapter 4). The legacy of conservative modernization is examined in Chapter 5, while the dimension of Brazil's crisis and the current dilemmas of Brazil and the world-economy are the subjects of Chapter 6. A short conclusion inquires how Brazil's challenges, as a semiperipheral country and a regional power, may affect the crisis/restructuring of the world capitalist system.

2

The incorporation of Brazil into the world-economy: from colony to national industrialization

During the distinct phases of Brazil's incorporation into the world-economy, its political economy has exhibited three patterns: that of colony, mercantile empire, and peripheral industrial capitalism. These three patterns extend from the original settlement by the Portuguese to the formation of the modern state and industrialization. Not until the late 1960s did Brazil emerge as a semiperiphery within the world-economy, becoming a regional power in the South Atlantic. The framework used here for examining the historical evolution of Brazil in relation to the world-economy is provided by Kondratieff's description of the world-economy's long waves, with modifications for internal conditions (see Table 2.1 on pages 18–19).

The colonial period (1530–1822) was characterized by the implantation of European mercantile firms based on slave labor and by large territorial expansion. It was followed by a long period of national slave mercantilism (1822–1889) under the political form of an empire, which was able to maintain diverse and fragile regional economies under a single authority. Empire, coffee, and slavery were the seminal marks of the formation of the Brazilian state. The era of national industrialization began at the end of the nineteenth century with a capitalist export economy which permitted the internal accumulation of capital and which strengthened the state apparatus, resulting in the national-developmentalism of the post-Second-World-War era. This chapter is devoted to describing these three processes. Brazil's new position as a regional power will be analyzed in a separate chapter.

The colonial period

The occupation and settlement of the territory which became Brazil is merely one episode in the wider process of maritime expansion which followed the development of European commercial enterprises. As a consequence of the Iberian countries' search for new routes to the Orient – Spain via the western route and Portugal by rounding Africa – the territory which today constitutes Brazil preceded the creation of the colony itself. The Treaty of Tordesillas, signed by the two countries in 1494, divided all the lands to be discovered between the Portuguese and Spanish Crowns. It established that all lands lying to the east of the 50th meridian west would belong to Portugal.

The treaty thus defined the colony, a priori, as a territory corresponding to only 40 percent of Brazil's current area but which was, even so, immense. The defense of this territory and its expansion did not result from military conquest. The process of possession was slow and complex, influenced by Portuguese strategy and favored by the fight for hegemonic power between the Dutch, French, and English, and by the union with Spain between 1580 and 1640.

Mercantile undertakings and defense of the Atlantic coast

Initially, the Portuguese commercialized dyewoods – brazilwood, for example, which later gave its name to the new colony – and hides prepared by Indians in modest trading posts along the coast.

The colonization of Brazil came as an afterthought to the Portuguese rulers, due to Dutch, English, and French pressure on the territory and following the loss of most of Portugal's Asian and African trading posts to the Dutch. The native population of the territory was relatively sparse in contrast to the Spanish territories. The Portuguese therefore could not rely on native labor, and initially no precious metals were found. It was thus necessary to organize the production of staples, and sugarcane plantations became the basis for the colony's economy and defense. This undertaking, until then unprecedented, owed a great deal to Portugal's prior experience on the islands of São Tomé and Madeira – encouraging the manufacture of equipment for sugar mills. It also owed a great deal to the commercial organization of the Dutch who controlled a large part of the sugar market in continental Europe.

Colonial Brazil was thus organized as a commercial enterprise resulting from the alliance of the Crown, a mercantile bourgeoisie

Table 2.1. *A space–time information matrix*

	Core
Logistic curve	
A *c.* 1450–*c.* 1600	Initial geographical expansion based on Iberia but economic advances based on northwest Europe
B *c.* 1600–*c.* 1750	Consolidation of northwest European dominance, first Dutch and then French–English rivalry.
Kondratieff waves	
I A 1780/90–1810/17	Industrial Revolution in Britain, "national" revolution in France. Defeat of France.
B 1810/17–1844/51	Consolidation of British economic leadership. Origins of socialism in Britain and France.
II A 1844/51–1870/75	Britain as the "workshop of the world" in an era of free trade.
B 1870/75–1890/96	Decline of Britain relative to United States and Germany. Emergence of the Socialist Second International.
III A 1890/96–1914/20	Consolidation of German and US economic leadership. Arms race.
B 1914/20–1940/45	Defeat of Germany, British Empire saved. US economic leadership confirmed.
IV A 1940/45–1967/73	United States as the greatest power in the world both militarily and economically. New era of free trade.
B 1967/73–?	Decline of United States relative to Europe and Japan. Nuclear arms race.

Source: Adapted from Taylor 1985: 20–21.

Semiperiphery	Periphery: Latin America and Brazil
Relative decline of cities of central and Mediterranean Europe.	Iberian empires in "New World." Portuguese colonization in Brazil, slave sugarcane plantations.
Declining areas now include Iberia, joined by rising groups in Sweden, Prussia and northeast United States.	Retrenchment in Latin America, rise of Caribbean sugar. Strengthening of Portuguese control, gold rush in Brazil.
Relative decline of whole semiperiphery. Establishment of United States.	Decolonization and informal controls in Latin America. Ports opening in Brazil.
Beginning of selective rise in North America and central Europe.	Expansion of British influence in Latin America. Independence and Brazilian empire, rise of coffee-based mercantile slave economy.
Reorganization of semiperiphery; civil war in United States, unification of Germany and Italy, entry of Russia.	The classic era of "informal imperialism" with growth of Latin America. Expansion of coffee plantations and crisis of slave labor in Brazil.
Decline of Russia and Mediterranean Europe.	The classical age of imperialism. Abolition of slavery and the Republic, capitalist coffee growth with immigration in Brazil.
Entry of Japan and Dominion states.	US hegemony in Latin America. Crisis of the coffee economy and early industrial growth in Brazil.
Socialist victory in Russia – establishment of USSR. Entry of Argentina.	Import-substitution in Latin America. Centralized power and beginnings of state planning in Brazil.
Rise of eastern Europe and "Cold War." Entry of OPEC.	Economic growth in Latin America. Rise of tripod-based industrialization and crisis of national-developmentalism in Brazil.
Entry of "little Japans" in East Asia and new regional powers – China, Brazil, Mexico, India. Rise of debts to core.	–

(including the Dutch), and the nobility. At the beginning of colonization, the legislation relating to landed property was based on the political system of rural Portugal. The land was viewed as part of the King's personal patrimony, as the Crown's dominion, and its purchase had the character of a personal grant according to the merits of the applicants and to the services they had provided to the Crown.

A strategy of controlled distribution of land relied on private entrepreneurs to colonize the territory without onus to the Crown, assuring the territory's occupation and control of the eastern shoreline. By geometric division of the Atlantic coast into hereditary captaincies (1530), colonization was initiated simultaneously in various points of the territory. Land was granted to recipients with the objective of promoting agriculture, above all the cultivation of sugar. The grantees had sovereign rights and could repartition the land among residents capable of exploiting it. These divisions respected the line set by the Treaty of Tordesillas even though the interior boundaries of the captaincies were unknown.

The next problem to present itself was that of labor and the indigenous peoples. They became the focus of an ambiguous policy, given the conflict between the Crown's interests in Christianizing the indigenous peoples and integrating them into the settlement of the territory, and the colonists' interests in enslaving them. The *Carta Regia* of 1570 established that the indigenous peoples could only be imprisoned in "just wars" and so, faced with the difficulty of obtaining labor, the colonists turned to the African slave trade, financed in large part by the Dutch.

As only part of the territory was utilized for commercial ends, landowners were able to maintain tenants and sharecroppers who would reside on the less fertile sections of their property. These small farmers dedicated themselves to a subsistence economy and worked on the plantation as needed. In this way, although profits were the principal motive for the economy, control over slaves, free people, and land were more important for defining the social status of landowners than was the accumulation of their riches (Viotti da Costa 1977).

The development of other economic sectors did not cause modifications in agrarian politics or in the typical labor arrangements in the sugar-growing areas. The assumptions which guided these practices in the sixteenth century survived into the nineteenth century. Although this strategy did not bring the desired economic prosperity, on the other hand it did establish the basis of the colony's economic, social, and political structure, and provided for continued control of the people and the effective occupation of the territory against external threats.

The coastal plantations were the basic cells of the economic and social structure of the colony. Cattle ranches gradually expanded through the backlands, radiating from these centers, and supplying the sugar-growing zones with leather and draft animals. Along the northern shore, the Amazon River was strategic because of its size and its ample navigability up to 2,000 kilometers into the interior through tropical forest. The Dutch, French, and English tried to occupy this area militarily during the union of the Portuguese and Spanish Crowns (1580–1640). The initial pattern of occupation to defend the Amazon Valley relied on small forts – the first one being at the mouth of the Amazon in Belém (1616).

To assure long-term occupation as well as the pacification and allegiance of indigenous tribes against the Dutch, English, and French, the Portuguese resolved to divide the valley among Catholic religious orders. They followed the example of the Spanish Jesuits who had already established a virtually uninterrupted strategic chain of missions in the heart of the continent, from the Prata River to the upper reaches of the Amazon during the sixteenth century and the first half of the seventeenth century (Prado, Jr. 1945a).

Territorial expansion beyond Tordesillas

In the century following the separation of the two Crowns in 1640, Portuguese colonization invaded areas which had belonged to Spain and occupied the territory which today is Brazil. Breaking the boundary set by the Treaty of Tordesillas became a goal for the metropolis and not merely a consequence of territorial defense.

The expulsion of the Dutch from the northeast, where they had remained from 1630 to 1654, brought with it the end of the Portuguese monopoly on sugarcane, as the Dutch moved to the Antilles and developed sugar cultivation there. With the destruction of its navy and the loss of its colonies in the East, Portugal became a secondary power, largely dependent on England to support it in the international sphere. Brazil then became Portugal's last overseas possession of great value, and the extension and control of the colonial territory was therefore decisive for the economic recuperation and affirmation of the centralized Portuguese state.

The occupation of territory as the basis for claiming possession – i.e. *de facto* possession – became the basic strategic practice for appropriating territory beyond the juridical limits of the Treaty of Tordesillas, hitherto recognized as the ruling legal principle. The practice of occupation was conducted in various ways, and concentrated in the interior and in the Amazon and Prata basins which were strategic for their navigability and for their positions at the extremes of the colony.

The greatest impulse to territorial expansion was the discovery of gold (1690) in the highlands of central Brazil. This became the colony's economic base until the middle of the eighteenth century, as the sugar economy declined in the face of competition from the Antilles. The discovery of gold provoked a flow of immigrants from the metropolis, large internal migration, and a giant "rush" for several decades. It covered an immense area in the center and west of Brazil's current territory (Minas Gerais, Goiás, and Mato Grosso). Roads for cattle and mule teams were established to supply the mining centers and constituted the first axes for the colony's internal integration.

As a consequence of mining, the economic axis shifted to the center-south, and in 1763 the capital was transferred from Bahia to Rio de Janeiro. However, the cycle of gold and diamonds, though intense, was brief. It ended in the last quarter of the eighteenth century, in part from the pressure of taxes levied by the Crown (one-fifth) which led to the first, but unsuccessful, movement for independence in Minas Gerais.

In the Amazon Valley, the Crown stimulated the activity of the missions. They became the greatest exporters of spices (cinnamon, cloves, sarsaparilla, native cocoa), in addition to producing food to provide subsistence for and to maintain the monopoly over indigenous labor. Forts and missions extended deeply into the Amazon region, assuring the future sovereignty of Portugal over an immense area, even if on a weak economic and sparsely populated basis (Figure 2.1).

In the extreme south at the end of the seventeenth century, a large vacuum of power existed between the Spanish, based on Buenos Aires at the mouth of the Prata River, and the Portuguese, whose occupation extended as far as the 26th parallel South. The Portuguese strategy had two facets. The aggressive side consisted in implanting a military garrison on the northern bank of the Prata River, just opposite the port of Buenos Aires, creating the colony of Sacramento (1689). This led to more than a century of war, encouraged by the English who sought to control the markets of silver, leather, and cattle in the Prata Basin. The pacific side of the

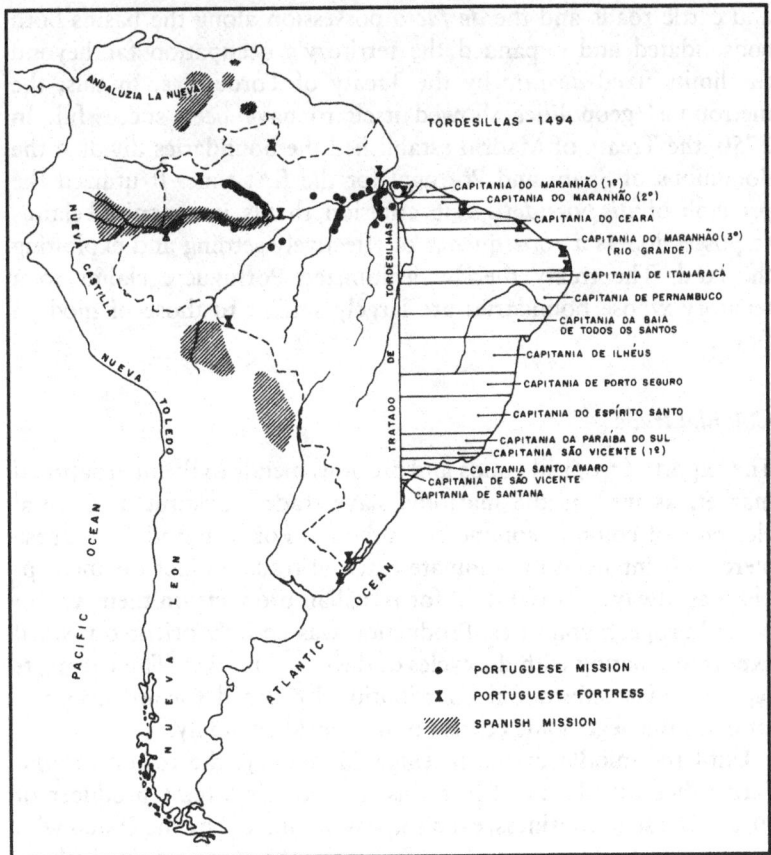

Figure 2.1 Territorial control in the colonial period. Adapted from Albuquerque, Reis, and Delgado de Carvalho 1980.

Portuguese strategy consisted of colonization directed by the metropolis which established some 4,000 Azorean couples near Porto Alegre and in Santa Catarina (1747). After peace was made in 1777, lands were distributed on a large scale to soldiers and cavalry in what is now Rio Grande do Sul, as a way of consolidating Portuguese possession. This was the origin of large pastoral latifundios, the *estancias*. Portuguese sovereignty was thus established simultaneously with the economic base of the region, which was already exporting dried meat to Rio de Janeiro and to Havana in 1780.

The rapid movement of mining and the slow expansion of ranches and cattle roads and the *de facto* possession along the basins both consolidated and expanded the territory's occupation far beyond the limits fixed *de jure* by the Treaty of Tordesillas. In this, the metropolis' geopolitics showed itself to have been successful. In 1750, the Treaty of Madrid established the boundaries dividing the dominions of Spain and Portugal for the first time. It utilized the principle of *utis possidetis* as its criterion, that is, it recognized claims to possession as a consequence of effectively settling and exploiting the land. The treaty thereby legitimized Portuguese claims to a territory whose boundaries are largely similar to those of modern Brazil.

Colonial trade

The export of tropical goods and precious metals to the international market, as well as the maritime slave trade, constituted essential elements of colonial commerce. Although Portugal and Portuguese merchants intended to dominate external trade, in fact the metropolis was always a waystation for Brazilian products on their way to other European countries. Production was entirely oriented toward export, oscillating with the cycles of the world market. These exports represented a substantial contribution by Brazil toward accumulation in the hegemonic centers of the world-economy.

Until the middle of the seventeenth century, the region around Pernambuco in the northeast was the world's largest producer of sugar. The sugar business, even so, was dominated by the Dutch who controlled its transportation: receiving the product in Lisbon, refining it, and marketing it in Europe. In the second half of the seventeenth century, with the rise of sugar production in the Antilles, sugarcane plantations in Brazil entered into a long crisis which extended until the nineteenth century.

During the seventeenth century, Brazil was also the world's largest producer of gold. Gold exports peaked around 1760 when they reached the sum of £2.5 million, but rapidly declined thereafter, falling below £1 million in 1780 (Furtado 1959). The cycle of gold contributed fundamentally to the development of manufacturing and finance in England, since Portugal was merely an extension, although an essential one, of the expansion of the English economy (Manchester 1973). Commerce agreements established privileges for Great Britain which in return agreed to defend the Portuguese possessions. The Treaty of Methuen in 1703 gave continuity to this alliance, allowing Great Britain to control the gold market, which

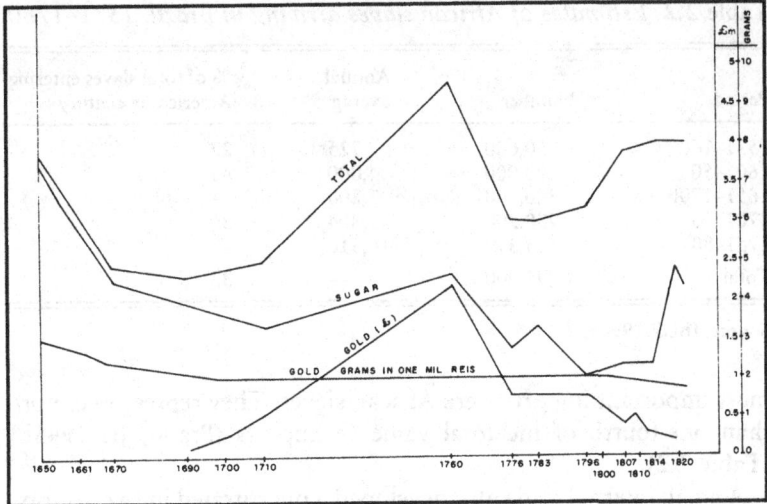

Figure 2.2 Brazilian colonial exports, 1650–1820.
Source: Simonsen 1937.

brought London as much as £50,000 of gold per week. The economic
presence of Great Britain was also based on contraband, particularly
with silver in the Colony of Sacramento. Therefore, foreign trade
remained a legal monopoly of the Portuguese, but was consistently
dominated by the English through treaty and contraband.

The colony entered a period of economic crisis during phase A of
the first Kondratieff cycle. The gold crisis (1760–80) cut the colony's
import capacity almost in half, with exports falling from £5 million
in 1760 to £3 million per year at the end of the century (Figure 2.2).

The gold crisis was partially offset by the Industrial Revolution,
the War of Independence in the United States, the French Revolution,
and the Napoleonic Wars which temporarily elevated export prices
and particularly favored the cotton and rice exports from Maranhão.
These products were introduced by trade companies created by the
Marquis of Pombal in 1755 and 1759. In 1807, Brazil figured as the
fifth most important source of cotton for England, to which it
exported 3,188,808 lb. Two other products had an important role
in the colony's exports in exchange for black slaves: tobacco and
sugarcane liquor.

As for the colony's imports, they consisted of various kinds of
luxury foods (wine and olive oil), salt, flour, and above all manufactured goods and processed metals – iron in particular. By far the

Table 2.2. Estimates of African slaves arriving in Brazil, 1531–1780

Period	Number	Annual average	% of total slaves entering America by century
1531–1600	50,000	725	22
1601–50	200,000	4,000	43
1651–1700	360,000	7,200	—
1701–50	790,200	15,804	30
1751–80	495,300	16,510	—
Total	1,895,000	—	33

Source: IBGE 1990a.

most important imports were African slaves. They represented more than one-fourth of the total value of imports (Prado, Jr. 1945b) (Table 2.2).

A small internal trade also developed, concentrated in two sectors. One sector, strictly dependent on foreign commerce, redistributed local and imported merchandise. The other sector provided the large coastal urban centers with supplies of cattle transported over land, and agricultural products transported by coastal shipping which connected the dispersed nuclei along the coast. External and internal trade were essential for the development of the urban mercantile bourgeoisie which included small Luso-Brazilian merchants, large merchants, and Portuguese traders. They also encouraged the emergence of local interests – the germ of a national interest.

Land, social, and spatial structure

The large-scale monoculture estates using slave labor were the basic cells of the entire colonial structure. They determined the entire pattern of social relations and the extensive and speculative exploitation of the colony's natural resources. Reproducing itself through time and space, the large estate's key features were most apparent in the cultivation of sugar. These features, however, were also present in cattle ranching, mining, and even in the extractive sector of the Amazon Valley – although there it was not based on the actual ownership of land.

Two basic units of production existed in the colony: the *engenho* and the *curral*. The *engenho* was the sugar mill that became the central element for organizing the plantations, eventually coming to designate the entire property. It was a complex, almost self-sufficient, establishment. The *engenho* normally used 80 to 100 slaves, but

Table 2.3. *Estimates of colonial Brazilian population, 1798*

Condition	1798
Whites	1,010,000
Free blacks	406,000
Indians	250,000
Total free population	1,666,000
Black slaves	1,361,000
Mulatto slaves	221,000
Total slaves	1,582,000
Total	3,248,000

Source: Conrad 1972.

some came to possess more than 1,000. The *curral* was a cattle ranch, producing the colony's only important non-export good. The *curral* extended over enormous tracts in the rural areas with undefined boundaries. It occupied lands inappropriate for cultivation and established a fundamental separation between agriculture and ranching. Cattle herds, varying from 200 to 1,000 head, were raised in the open by a ranch worker with two to four helpers.

The principal characteristic of colonial society was the enslavement of indigenous peoples and, even more, of Africans. It was an original form of slavery, reviving a form of labor which had been historically extinct and providing a means for commercially exploiting the colony. Although complete and reliable data are lacking, the share of slaves in the total population in 1798 has been estimated at close to one-half (Table 2.3).

Slavery affected the very concept of labor, which became a dishonorable activity, and left only a small margin of employment for free individuals who were not landowners. Two sectors of society thereby came to be distinguished in colonial society, based on whether they utilized slave or free labor. One sector was organized around slavery and the cohesive patriarchal clan formed by groups of individuals who participated in the large rural estates. Owners of these estates constituted a privileged class, an aristocracy. Only merchants had the power to confront these landowners, as financiers for large-scale cultivation. The rivalry between these two groups was accentuated by the fact that the merchants had been raised in Portugal while the rural landowners were generally born in the colony.

The other sector was without structure, encompassing the mass of the free population squeezed between the two extremes of the social scale, the landowners and the slaves (Franco 1974). They were "poor whites," "civilized Indians," and free blacks with uncertain and sporadic employment, or without any occupation at all. In every declining phase of an economic boom, the part of society affected by the crisis was dislocated. This created a residual social category made up of individuals without fixed employment or roots – highly mobile and fluctuating about the organized segment of society.

The economic and social structure of agro-exports was associated with a specific spatial structure. At the end of the colonial period, at the beginning of the nineteenth century, a population of 3.5 million inhabitants was distributed irregularly across the vast territory. This pattern persisted in large part until the middle of the twentieth century. The first and most marked aspect of the spatial structure was the profound difference between the *marinha* (marinelands) and the *sertão* (backlands). Some 60 percent of the population was concentrated in the coastal band, about twenty kilometers in width, while the rest was dispersed throughout the backlands. The coastal regions contained "civilization": sugar plantations, cities, and ports. The backlands contained a primitive society and a sparse, dispersed population which made any kind of control difficult. This nebula of isolated settlements was spread through an area of more than 2 million square kilometers.

A second aspect of the spatial structure was the concentration of production and social organization within the coastal band itself. The isolation of these nuclei – unconnected to one another and each with its own autonomous port oriented toward Europe and its separate system of river transportation – displayed a spatial structure best described as an "archipelago." The articulation of the whole was maintained by weak commercial connections in the internal market which, nonetheless, were capable of generating the first significant cities of the colony, particularly Rio de Janeiro.

The mercantile empire

In the nineteenth century, Brazil became part of industrial capitalism as a politically independent state, in the form of an empire that lasted from 1822 until 1889. Brazilian independence was actually part of Britain's strategy as a new world power to control all of South America through "free" trade. Industrial capitalism generated contradictions with the colonial pacts which were based on the metropolitan monopoly over trade and whose destruction was a necessary condition to allow the periphery to achieve mass production and constitute a market for industrial goods. The rich Iberian colonial empires, until then protected by the rivalry between England and France, fell apart with the Napoleonic Wars. The obstacles for the new international order were therefore removed, and the free Ibero-American nations emerged. In the second half of the nineteenth century, these nations were incorporated into the world-economy as part of informal imperialism.

With the birth of the nation-state, Latin American history ceased to be merely reflexive. Domestic decisions began to influence the form of Brazil's incorporation into the world-economy. This was followed by a long era of coffee-based mercantile slave economy.

Formation of the national mercantile coffee-based slave economy (1790–1850)

At the time of its independence Brazil was the largest country in the West, yet it had hardly 4 million inhabitants. Why did the former Portuguese colony not fracture politically at the moment of independence, as occurred with the formation of the Spanish American republics? And why was slavery not abolished? Because of specific internal factors, basically the alliance forged between local power-holders, particularly coffee-growers, and merchants. To maintain their privileges – slavery and monopoly of land – they favored independence in the form of a monarchy and territorial unity. Thus, slave labor was maintained because at that moment it in no way affected the entry of British products and because its transformation into free labor had become a domestic issue.

Independence and politico-territorial unity

Brazil's colonial era ended in 1808, even before its independence. The "opening" of Brazilian ports to the international market, after the Portuguese Court transferred to Brazil in flight from Napoleon's armies, marked a new economic period which unfolded under British hegemony. The British strategy was clear: having lost Portugal to France, it sought compensation in the large Portuguese colony of Brazil. The Portuguese sovereigns preserved their Crown and titles by sailing to Rio de Janeiro under the protection of a permanent British naval division, but lost their independence of action. With the opening of the ports in 1808, not only was the commercial monopoly broken, but Portuguese domination virtually ceased due to the

special favors conceded to Britain in the commercial treaty which
was signed in 1810.

The interests of the British commercial and industrial bourgeoisie
played an important role in Brazilian independence. To these were
added decisive local interests. The export crisis at the turn of the
century and the extortionate taxes established by the parasitic
metropolis provoked dissatisfaction among the plantation owners
and merchants, as well as social agitation. The presence of the court
in the colony, the opening to the international market, and the
elevation of Brazil to the status of a vice-kingdom (1815) stimulated
economic activities, urban growth, and the centralization of society
around the power center of Rio de Janeiro. The emigré bureaucracy
of the court acquired local interests and became sympathetic to the
liberation of the colony.

Conditions were thus created for independence, which was
declared by the heir to the Portuguese throne. These conditions by
themselves, however, do not explain the continuation of political and
territorial unity, a controversial theme among historians even today.

Preserving this unity was in Britain's interest because the new
country represented a large potential market and a base for operations in South America, particularly in the dispute with France over
controlling navigation and trade in silver and gold in the Prata Basin.
Even before, in 1810, Brazil had allied itself with England in the war
that resulted in the fragmentation of the vice-regency of the Prata
Basin against the interests of the French and their Spanish allies, in a
dispute which continued until 1852. But the British pressure against
the slave trade, which constituted an obstacle to their domination
over the South Atlantic, initiated a conflict between that country and
the dominant groups of the colony.

But it was the maintenance of the monarchical principle during the
process of independence that determined the preservation of the
political unity of the territory. The centralist monarchical principle
was the solution that the great landholders and slave traders found to
defend their privileges and maintain their local power: externally, to
guarantee the slave trade against British pressure; internally, to
guarantee the interprovincial commerce in slaves, and the monopoly
of landed property (Saes 1985). The dominant groups of the
different areas overcame their isolation and accepted the monarchic
form of state for the whole national territory. And certainly the
presence of the heir to the throne and the practice over time of
defending the possession of the territory also influenced the implantation of the monarchy and the maintenance of territorial unity.

This relation between the centralism of the state and slave interests reveals the character of the power bloc. It was composed of all the categories of property owners but was under the political hegemony of a sub-bloc, the elite of slaveholding coffee-planters, urban propertied groups, and merchants – these last acting as a vanguard and conquering key posts in the state apparatus. It reveals also the beginning of differentiated regional interests.

The separatist struggles that ensued were directed principally by members of non-slaveholding propertied groups, held at a distance by the power bloc. Between 1831 and 1848, twenty provincial movements broke out throughout almost the entire national territory. English officers commanded Brazilian troops in liquidating these separatist movements. The Constitution of 1824, which lasted until 1891, consecrated the dominance of the executive administration over the other branches of the state apparatus, and of the central sphere over the regional and local spheres – although in practice the local power of the great landholders continued.

The gestation of the coffee economy

The breaking of the Portuguese monopoly over trade and the formation of the nation-state marked the end of the colonial economy's crisis. However, the crisis' resolution was not predetermined. The economy could have stagnated but, to the contrary, its mercantile and slave character were revitalized, now in the guise of a nationally controlled economy. Coffee, the new product, came to dominate the national mercantile slave economy and to change the direction of the crisis.

The coffee economy was the work of national commercial capital which had gradually developed, receiving a notable impulse from the opening of the ports, the transfer of the royal court to Brazil, and, later, independence. Merchant capital captured the new opportunities by commercializing coffee in the international market. It invaded the orbit of production by financing the startup costs for coffee cultivation through a new actor, the "commissary," which substituted for the previous commercial monopoly with the same result (Stein 1961). Three other factors favored the coffee economy as business: the availability of land in the Paraíba Valley which was near the port of Rio de Janeiro, the exploitative cultivation of the land, and the extraordinary level of exploiting slaves.

It was thus possible for Brazil to produce coffee on a large scale and at low cost. Brazilian supply lowered international prices and expanded external demand (Mello 1982). Between 1821 and 1830, coffee already represented close to 20 percent of the value of Brazilian exports and from this period until 1850 coffee exports increased fivefold (Furtado 1959). By 1830, then, Brazil had become the world's primary producer of coffee – and coffee had become the single largest export for Brazil and South America (Simonsen 1937).

At the same time, imports grew, mainly through increases in the slave trade. Fifty thousand enslaved Africans were imported each year, valued by the market at approximately £8 million (Prado, Jr. 1945a). The commercial balance was chronically in deficit, and was only solved in the medium term by extremely onerous British loans. By 1852, there were already four major loans valued at £2.5 million, generating a permanent budget deficit.

Suspension of the slave trade and the Land Law

While they could, the dominant slaveholding groups held back the abolition of the slave trade, which was only decreed under British diplomatic pressure in 1850. That same year, new policies were established to regularize the possession of land and the supply of labor, as well as credit: the Land Law.

Under this law, land became public domain, the nation's inheritance, and could only be acquired through direct purchase from the government. This eliminated traditional forms of acquiring lands through occupation ("possession") and through royal grants by the Crown. Lands that were not adequately utilized or occupied had to be returned to the state as public lands – "returned lands" – whose sale provided funds for their demarcation and for subsidizing the immigration of "free" European colonists.

Thus the expansion of the traditional regime of territorial property was blocked, as was the development of small independent production, including that of immigrants who had no financial resources to purchase land. The Land Law, therefore, assured the maintenance of the immigrants as laborers in the coffee plantations.

Even after this new law, occupation continued to be the principal means of access to land, but only for the large landowners who had the necessary resources. Land became both a commodity and a source of economic power. The export–import slaveholding society, then, was perpetuated at the same historical moment at which the Homestead Act increased land access for smallholders in the United States (Viotti da Costa 1977).

The state thus fulfilled its role with the slaveholding bloc. This made the government a mediator between the public domain and the probable property-holders through an impersonal relation, giving it an instrument of symbolic power. The initial challenge to consolidate a power center in a vast territory – lacking regional integration and sparsely populated – made "stateness," based upon alliances with private landowners, much more relevant than "nationality." The bureaucratic power that developed from the 1850s was mainly a symbolic affirmation of a public order, since in practice the population – slaves and free whites – continued to owe their primary allegiance to local private landholders (Reis 1983).

The apogee of the mercantile economy and the decline of servile labor (1850–1888/1989)

In the middle of the nineteenth century, world capitalism under England's leadership stimulated the export of capital and production goods, constituting the classical era of "informal imperialism." Not all the Latin American states were recipients of capital exports, depending on their levels of internal economic organization. Brazil and Argentina absorbed 60 percent of British investments in Latin America (Silva 1976). The Brazilian state, endowed with the capacity to sustain an export economy including guarantees to foreigners for interest and loans, permitted new British capital flows once the conflict over abolishing the slave trade was ended.

Once more, Brazil linked itself to the technologically advanced power of the era, Britain. It favored the utilization of nineteenth-century technological progress, enabling a surge in the export economy, which had been impossible during the colonial period. The initial phase of the crisis following the suspension of the slave trade was overcome. The coffee economy advanced, dominating the country's production until the first quarter of the twentieth century.

Imperial geopolitics

The Brazilian imperial government's difficult situation did not mean that it had lost sight of the importance of territorial control. At the time of independence, Brazil's boundaries were neither defined nor marked, and their definition came about largely as a consequence of the dispute between England and France over control of navigation and commerce in the two grand basins which demarcate Brazilian territory, the Prata Basin and the Amazon Basin. But the state itself showed growing interest in these areas in its geopolitics toward the two basins.

English and French interests collided in the Prata Basin as early as the seventeenth century. Each was allied, respectively, with the Portuguese and the Spanish for control of navigation and trade in silver and leather. In the seventeenth century, Brazilian interests in navigation on the Prata River expanded because the river provided links to its distant western province, Mato Grosso, settled during the gold boom. The conflicts over borders with Argentina resulted in the creation of Uruguay (1828) as an independent republic between the two rival forces.

Directed colonization, the form of controlling territory in the frontier areas used since the colonial period, was the imperial government's tool for defending its two meridional provinces which, isolated by extensive forests, were threatened from the south by the Argentines and from the interior by the indigenous peoples. The new colonists had to be as much soldiers as farmers, in order to defend and cultivate the land. Many of the new colonists were soldiers released from Napoleon's armies and poor peasants from Central Europe, principally from Germany, who became small family landholders (Waibel 1958). Begun in the 1820s, immediately after independence, foreign colonization remained insignificant until the extinction of the slave trade in 1850, and expanded rapidly thereafter.

The policy of economic autarchy and independence adopted by Paraguay, a landlocked country dependent on the port of Buenos Aires, contradicted the interests of both Brazil and Argentina. When the Paraguayan government pressured Uruguay to grant it access to the sea, the greatest war ever unleashed on the South American continent broke out (1864–68). Brazil, Argentina, and Uruguay allied themselves against Paraguay, with English encouragement and finance. Brazil was victorious, but at the cost of 140,000 wounded and 33,000 dead. Brazil was also financially drained by the war. Paraguay was mutilated, losing almost all of its male population in the war, and was never able to reassert itself. The bloody victory was Brazil's final thrust to define its southern border.

Paradoxically, the imperial policy toward the vast and distant Amazon Basin was the opposite of that employed in the Prata Basin. Brazil tried to block free navigation and external incursions, and it was able to retain the area because of a dispute between foreign powers. In the eighteenth century, England favored

Portuguese claims to the Amazon in order to exclude France. In the second half of the nineteenth century, the development of the rubber economy brought US interests into the area. Because of rubber exports, the region came to be integrated with the international market before it was integrated with the rest of Brazil. Albeit with token resistance on the part of the central government, in 1867 the Amazon was opened to free navigation.

The end of slavery and of the empire

In the middle of the century, coffee production and the country's wealth were concentrated in the middle and upper reaches of the Paraíba Valley. The prices in the international market, despite oscillations, maintained the expansion of production which, even so, was limited by the growing costs of transportation and slaves. At the end of the 1860s, economic crisis broke out in the Paraíba Valley, already in decay due to the extreme exhaustion of the land.

The coffee economy did not regress, however, for three reasons. First, the surge of railway construction at the end of the 1860s lowered transportation costs. Railway construction was made possible by the interlacing of national merchant capital, English financial capital (the principal owner or financier of the railroads), and the state (which guaranteed the interest payments on the investments). The second factor was the introduction of machinery in the 1870s which benefited coffee production in western São Paulo. This machinery saved labor and improved the product while lowering prices. Third, the exceptional conditions of high plains and fertile purple soils created by the decomposition of basalt in the São Paulo highlands permitted coffee production to expand, forming a veritable "sea of coffee" (Monbeig 1952). Between 1870 and 1880, the number of coffee trees increased from 106 million to 220 million. Rapidly occupying an almost unpopulated area, coffee stimulated the growth of the city of São Paulo as well as the port of Santos which began to compete with Rio de Janeiro (Geiger 1963). By the middle of the nineteenth century, coffee was the chief Brazilian export good, and the United States absorbed more than 50 percent of it, revealing US dominance over the mercantile links (Figure 2.3).

With the coffee boom, the external-account balance was restored. From 1860, external trade showed growing surpluses despite the expansion of imports for the consumption of certain groups and regions thanks to new infrastructural apparatus (railroads, ports, communications) financed by British capital. Investments from

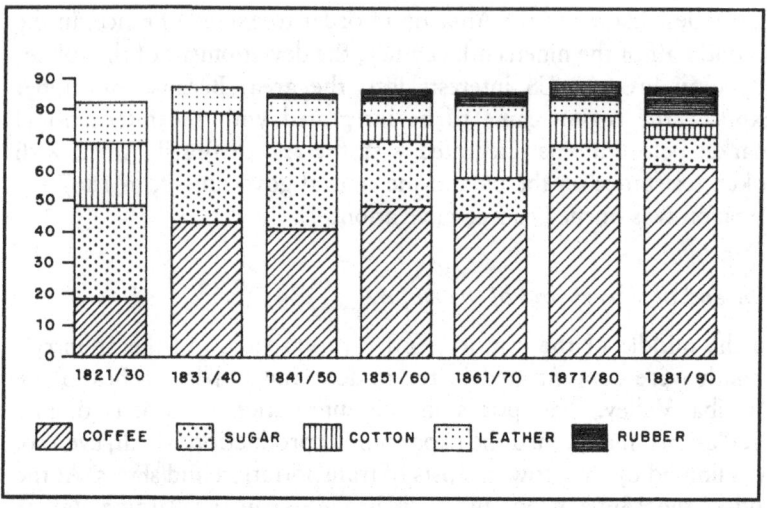

Figure 2.3 Brazilian exports during the empire, 1821–1890 (percentages). Basic data from IBGE 1990a.

Britain multiplied; however Argentina received the major share at the end of the century (Table 2.4).

The national mercantile coffee economy reinvigorated itself, but its expansion was increasingly constrained by the scarcity of labor in the new zones. The expansion of coffee required the formation of a labor market.

Slavery was abolished in 1888. In the following year the monarchy was overthrown and an economically liberal, republican form of government was installed. The interrelated movements created juridico-political conditions for the predominance of capitalist relations of production in the subsequent decades: free labor and the reorganization of the state apparatus according to principles of bourgeois bureaucratism, that is, formally open to all social classes (Saes 1985).

The expansion of the coffee economy intensified the problem of labor supply. By the middle of the nineteenth century, attracting foreign immigrants had already been attempted as a solution – through the system of *colonato* in which settlers took responsibility for a certain number of coffee trees, receiving half the net profit. They were indentured servants (Conrad 1972). After the coffee boom of the 1870s, to compete with the United States and Argentina in attracting European immigrants, the government of São Paulo began

Table 2.4. *Total British investments in Latin America and Brazil, 1825–1913 (in £ millions)*

Year	Latin America	Brazil	%
1825	24.6	4.0	16.2
1840	30.8	6.9	22.3
1865	80.9	20.3	25.1
1875	174.6	30.9	17.7
1885	246.6	47.6	19.3
1895	552.5	93.0	16.8
1905	688.3	122.9	17.9
1913	1,177.5	254.8	21.6

Source: Silva 1976.

to subsidize immigration in 1881; later, the federal government joined this effort as well.

The contradictions of the slave regime accentuated. Throughout the nineteenth century, slaves were drained from the northeast to the more economically dynamic south, exacerbating the conflict between distinct regional interests which fractured the dominant groups: planters and exporters of sugar in the northeast versus growers and commissaries of coffee in the southeast, whose interests were linked to those of England and the United States. They launched a struggle for provincial autonomy, for federalism, which would only be possible within a republic. They demanded no less than a change in the form of the state.

This change, though, came about through another, broader struggle which was impelled by a middle class also desiring to change the nature of the state. In the second half of the nineteenth century, an urban middle class emerged. It included small manufacturers as well as non-manual workers who lacked ownership of their means of production – office employees, bureaucrats, military personnel, and liberal professionals. By the 1872 census, out of a total population of 9,930,000 inhabitants, professionals were already a small but visible subgroup. The census recorded 2,600 religious, 2,000 doctors, 3,500 teachers, 7,000 law practitioners, 10,700 public employees, and 28,000 members of the military.

The ideology of the middle class sought the valorization of non-manual labor as a means of differentiation from slave labor, creating a hierarchy of workers based on their individual capacities. For this, it became necessary to extinguish the ideology of slavery which devalued labor in general. The middle class thus took the lead in transforming the state, organizing and spreading the slave revolt, and imprinting upon it the goal of reorganizing the state apparatus according to the principle of merit.

The monarchy was finally toppled by a military movement seeking to reorganize the state apparatus in order to participate in political power. The imperial Armed Forces were strengthened by the conflicts with neighboring nation-states, but the criteria for recruitment continued to be based on favoritism. The tendency toward pro-fessionalization, hierarchization, and the creation of objective rules for promotion impelled some of the middle-level officers of the Army to identify themselves with non-manual laborers and with their criticism of the devalorization of labor, opposing themselves thereby to the dominant groups and to the government. The military group went on to act as a true republican party organization, formalized with the creation of the Military Club in 1887 and between them the positivist ideology was dominant.

The proclamation of the Republic was thus via the military, expressing the interests of the nascent middle class and of the liberalism which was consecrated by the Constitution of 1891. After 1894, however, the dominant classes governed once more through an oligarchic pact. Within this group, the regional coffee bloc retained hegemony.

The state and national industrialization (1888/1889–1967)

The third and fourth Kondratieff cycles (1896–1967) corresponded to a process of globalization in the world-economy under US hegemony. Different parts of the planet, however, were incorporated in ways that varied according to each area's distinct concrete historical circumstances. Brazil maintained and strengthened its links to the world-economy by consolidating its national industrial capitalism with the growing participation of a state which came to exercise ever-increasing control over the internal market. The state intentionally encouraged a process of import-substitution, modifying the effects of the world market to favor the national elite, but in this process it achieved relative autonomy.

In the first half of the 1900s, the coffee agro-exporting complex continued to dominate the Brazilian economy. The state was strengthened by its interventions to protect the complex. After 1930, the state also began to plan industrial development. In addition, the consolidation of national territory became a growing symbolic resource for legitimation of the state.

After the Second World War, the United States acquired definitive influence in Latin America. Brazil positioned itself in the world-economy through a process of politically directed industrialization, in a strict partnership between foreign monopolistic capital, state capital, and private national capital – inaugurating the famous *tripe* (tripod-based) model.

The capitalist export economy and early industrial growth (1888/1889–1945)

Brazil was incorporated into the emerging monopolistic capitalism of the end of the nineteenth century as a capitalist export economy. In Latin America, imperialism based upon financial control directed itself toward the formation of export companies and the creation of the infrastructure they required to operate. It also promoted the mass immigration which was essential to the birth of the capitalist export economy.

In 1889, Brazil's population grew to 14 million inhabitants, still concentrated near the coast. Its commercial trade reached £50 million. In this period, Brazil's coastal region was already connected by 9,000 kilometers of railroad and 1,000 kilometers of telegraph wires. This strengthened the spatial pattern of an "archipelago," with a large concentration of infrastructure in the "island" of the coffee-growing region. The commercial and financial apparatus for the coffee economy, which constituted itself as an agro-exporting complex, generated a coffee-growing bourgeoisie. The internal market expanded with the emergence of urban middle classes and with the substitution of slaves by free laborers – who solved the labor-shortage problem and liberated capital for other investments. In sum, conditions were created for the emergence of industry.

After the political turmoil which came with the federalist regime, the First Republic (1889–1930) consolidated itself as a "Republic of colonels," i.e. landowners with enormous local power, under the hegemony of São Paulo and Minas Gerais. In the Vargas era which followed, stateness consolidated itself in an authoritarian form and government planning was initiated.

Consolidation of boundaries and internal territorial control

In the first half of the nineteenth century, frontier disputes had arisen from the clash of English and French interests. By the middle of that century, the empire had already come to exercise a national geopolitics for the Prata and Amazon Basins, revealing the territorial interests of the Brazilian state. In the First Republic, however, the significance of territory began to change; it became a resource to be used in the legitimation of the state by means of diplomatic and military activities.

The diplomatic strategy was essential for consolidating the northern boundaries. By the end of the nineteenth century, the chief diplomatic axis was displaced from London to Washington, and Brazilian diplomats saw a favorable moment to conduct diplomatic negotiations. The redefinition of Brazilian territory became an essential legitimation device and a precondition for national unity (Machado 1987). The series of negotiated settlements of national boundaries that started in 1891 involving the Amazon is one important case, considering the value that rubber had acquired in the country's trade balance.

Based once again on the principle of effective occupation, some hundreds of thousands of square kilometers were added to Brazilian territory. Boundary settlements with Great Britain (the Essequibo area), with France (the Orinoco area), and with Bolivia (the Acre area) were negotiated by Baron Rio Branco, the chief statesman promoting Brazilian interests. The advance of rubbertappers along the upper reaches of the tributaries on the southern margins of the Amazon came into confrontation with the Bolivians. An international military conflict ensued which ended when Bolivia ceded almost 200,000 square kilometers of territory, for which it received from Brazil indemnities of £2 million and the construction of the Madeira–Mamore railroad which would guarantee Bolivian access to navigation on the Amazon. In 1903, then, the Territory of Acre was formed under direct administration by the federal government. In all the negotiations of this period, the only decision unfavorable to Brazilian aims in the Amazon region was in the conflict with Great Britain over the Essequibo region.

Internal control of the territory was conducted by the military. The passage from monarchy to republic was marked not only by regional conflicts but also by movements which resisted the modernization of society even though maintaining its authoritarianism. These included all the movements of the backlands, large spaces uncontrolled by the central government which were the unlimited and arbitrary domain of the large landholders, local political bosses – the "colonels" – and their henchmen.

The movements originating in this context assumed a mythic and messianic character – making the backlands a "locus" for liberation from existing social relations, a promised land. They affirmed the other face of the backlands from before the abolition of slavery – that of refuge and center of slave resistance.

Most of the messianic movements, but not all, occurred in the backlands of the northeast. The most important was at Canudos. It was a virtual fortress of poor people that resisted the Brazilian Army's assaults for an entire year (1896–97). In the Contested backlands, a pioneer area disputed by the states of Santa Catarina and Rio Grande do Sul where the "colonels" strengthened themselves, a messianic movement lasted for more than thirty years (1882–1915), before being finally defeated by military force.

With military efforts focused upon internal control, they began to participate firmly in political power. The military gave priority to developing transportation and communication links – extending the telegraph network – so as to maintain control over the country's immense physical base. They sought to secure the frontiers and break the isolation of the backlands, which simultaneously signified the exploitation of the territory and the valorization of land in the interior. The Army understood that private international capital invested only in infrastructure for transport and urban services in the big coastal cities and that private national capital only undertook investments with assured short-term profits. Therefore, they took upon themselves the task of promoting territorial integration, based on their resources, including technology (CECSIB 1972).

Coffee crisis and the emergence of industry

The formation of the labor market was an essential condition for the capitalization of the economy. After the 1870s, European immigration was stimulated toward obtaining wage laborers for the expansion of coffee in São Paulo. On the other hand, the large influx of immigrants to Brazil was made possible at the end of the nineteenth century by the transformation of the European economies – which created conditions for the formation of an international labor market, along with the simultaneous economic stagnation in the United States and Argentina.

In 1886, 30,000 immigrants entered Brazil; 55,000 entered in 1887; 133,000 in 1888; and from then onwards some 100,000 immigrants entered each year, mostly to São Paulo (Figure 2.4). They were largely of southern European origin (90 percent), and almost 60 percent came from Italy (Graham 1972; Graham and Merick 1979).

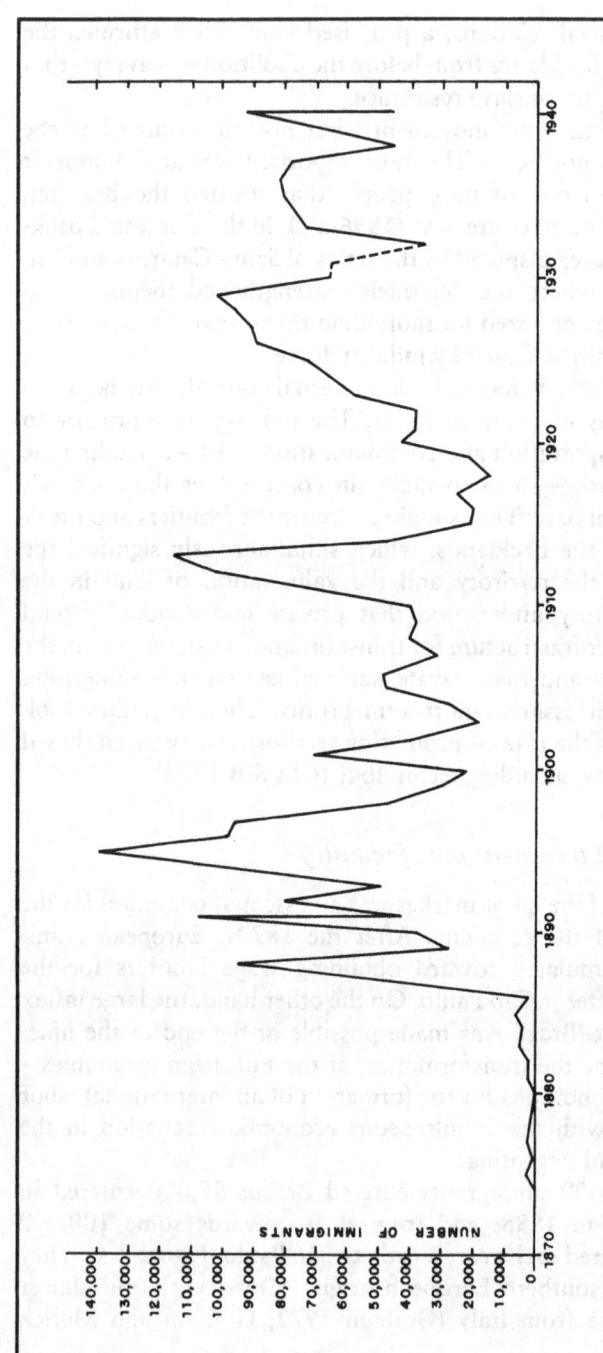

Figure 2.4 Immigration to the State of São Paulo, 1870–1940.
Source: Monbeig 1952.

In the coffee regions, the immigrants became wage laborers, with the assistance of government subsidies. In the forests of Brazil's three southern states, far from the large plantations, immigrant colonies multiplied under the stimulus of local government and the action of private companies. These small producers became not only an available labor supply but also the principal source of foodstuffs for the markets of Rio de Janeiro and São Paulo, to the degree that they were linked to these major cities by the railroads.

The pattern of primary exports diversified with the extraction of rubber in the Amazon and the cultivation of cacao in Bahia. The coffee-exporting complex, however, remained dominant, combining an agrarian and industrial producing sector with a commercial urban sector (Cano 1977). It expanded vigorously due to various conditions, including rising prices; ample availability of land, thanks to the extension of the railroads; an abundant supply of labor generated by massive immigration – for both the producing and urban commercial segments; and income transfers toward coffee capital promoted by the state. Faced with crises of overproduction, the government initially used currency devaluations to maintain the internal product price. Later, at the demand of the producers, the state began to purchase output in periods of crisis so as to maintain prices – the so-called valorization for coffee (Holloway 1978). The purchases were effected with resources obtained from financial groups who came to control the coffee trade (German, English, French, and North American). After 1924, the government responded to a period of surplus production by establishing a permanent support program, continually seeking to maintain prices at a high level.

Politically, the policy of supporting the coffee-planters meant strengthening authoritarianism. Although political ideology was vague, São Paulo's coffee-planters moved from economic liberalism toward a complete adhesion to market regulation. This ideology, once dominant, provided crucial legitimacy to the expansion of public power, even though within an oligarchic context. State-building went far ahead of nation-building, for there was no corresponding expansion of the political arena in order to incorporate extra-agrarian interests during the First Republic (Reis 1983).

Several revolts occurred against the oligarchic pact. The most important was the Army lieutenants' movement which gave origin to the "Coluna Prestes," a military group commanded by Luis Carlos Prestes – later leader of the Communist Party. The "Coluna" crossed the country during several years trying to organize the people against

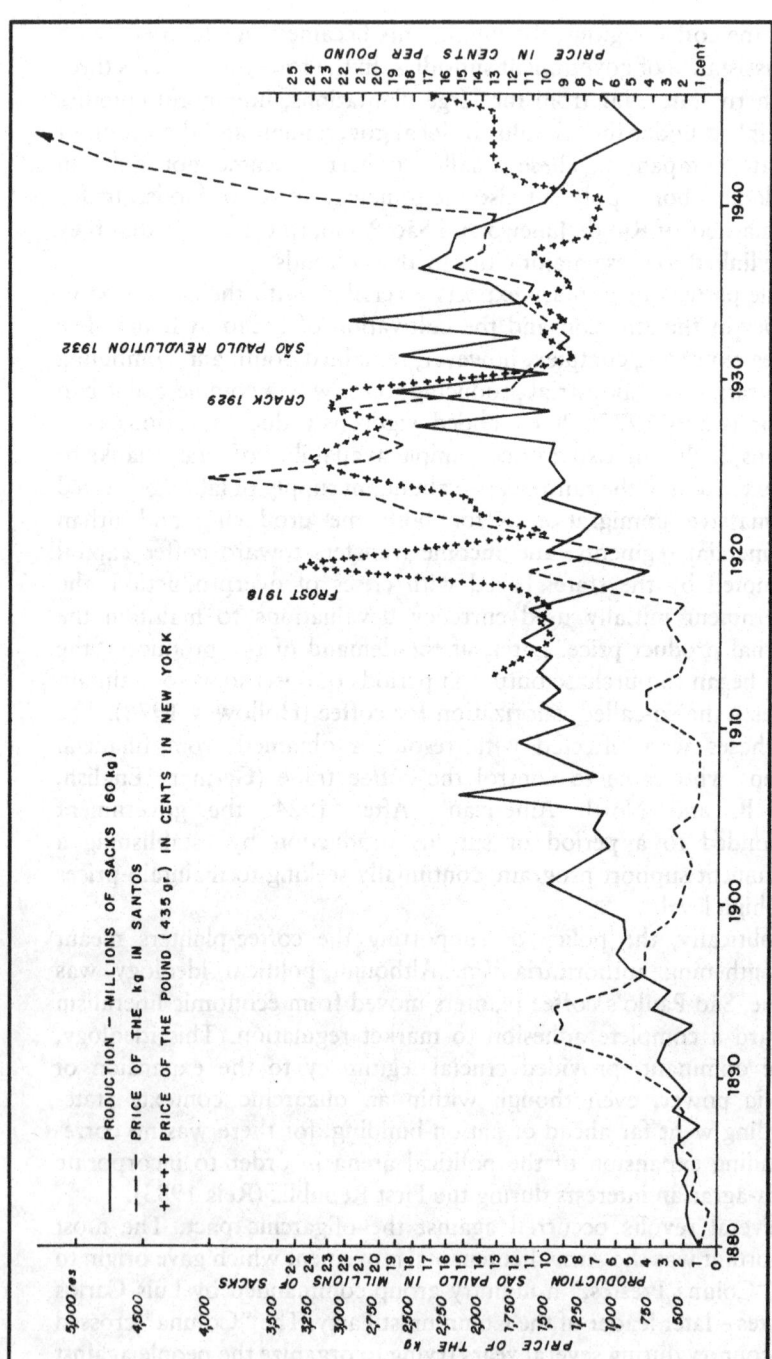

Figure 2.5 Coffee production and prices, 1880–1946.
Source: Monbeig 1952.

the "colonels." Italian immigrants, belonging to the anarchist movement, organized the first labor strikes in São Paulo, which were severely repressed and led to banishment of many working-class leaders.

The enormous production of coffee between 1924 and 1930 (the number of trees expanded from 940 million to 1,155 million), exacerbated by the production of competitors, especially Colombia, required the purchase of considerable stocks by the government. Consequently, the government assumed a growing debt. In the face of the Great Depression of 1929, the coffee-exporting complex entered into a crisis whose effects extended throughout the national economy. The price of coffee on the New York market fell from 23 cents per pound to 8 cents, bringing with it the ruin of many producers who, already indebted, lost their lands to the banks (Figure 2.5). The political consequence of the crisis was the Revolution of 1930, when Getulio Vargas took power, aided by the military.

Nevertheless, the coffee-exporting complex created various conditions for the emergence of a process of import-substituting industrialization. These conditions included the demand for wage goods in the producing areas and among the urban sectors; the existence of surplus commercial profits, for the coffee bourgeoisie; the state's actions in conceding easy credit to agriculture, permitting the banks to turn themselves into industrial entrepreneurs, and establishing protective tariffs; the supply of labor in the cities; the capacity to import production goods, as well as food and consumer goods necessary for the reproduction of the workforce.

Therefore, the consumer-goods industries began to expand, especially the textile industry. The first general census of Brazilian industries in 1907 registered 3,258 establishments which employed 150,841 workers. Among these, textile firms were foremost (some 60 percent of the total) followed by food processing (15 percent). Industry was spatially concentrated in Rio de Janeiro (33 percent), São Paulo (16 percent), and Rio Grande do Sul (15 percent).

Above all, industry was made up of firms of private individuals, a large number of them belonging to foreign immigrants. After the First World War, given the difficulties of importation, large foreign assembly companies established subsidiaries in Brazil – as in the case of the chemical and pharmaceutical industries, motor vehicles produced by General Motors and Ford, or the meat-freezing firms of Swift, Wilson, Armour, and Anglo.

In the census of 1920, the number of industrial establishments had grown to 13,336 with 275,512 workers. In the 1920s, the industrial structure was further diversified by the addition of a small steel industry, beginning with the installation of the Companhia Siderurgica Belgo-Mineira and others, and the implantation of the cement industry with the Companhia Brasileira de Cimento Portland (with Canadian and US capital).

Centralized power and the beginning of state planning: the Vargas era

The intense development of coffee capitalism created the conditions for its own demise by engendering the fundamental prerequisites for the Brazilian economy to respond creatively to the crisis of 1929. In this, the state carried out an economic program which promoted rapid economic recuperation in the midst of the recessive phase of the third Kondratieff cycle.

The collapse of the international economic order in 1929 widened the divergence between the great landowners, the coffee interests, and the central government. State intervention eliminated archaic rural structures and promoted industrial development. The Revolution of 1930 brought Getulio Vargas to power, and he continued as the country's dictator until 1945. This marked the end of the Old Republic and the consolidation of a modern state apparatus with an authoritarian character, the "Estado Novo," formally implanted with the coup of 1937.

After 1930, accumulation followed a new pattern: endogenous industrial expansion with an urban base. Between 1931 and 1937, the federal government sustained the nation's import capacity by removing surplus coffee from the international market with the purchase and destruction of coffee stocks, strongly protecting the small capital-goods industry, and blocking investment in the consumer-goods sector by prohibiting the import of new equipment. Between 1929 and 1939, industrial growth attained an average rate of 8.4 percent against 2.2 percent for agriculture (Villela and Suzigan 1975). But this industrialization found itself "restricted" because the technical and financial bases for accumulation were insufficient to implant, at a single blow, the basic nucleus of a capital-goods industry (Tavares 1975; Mello 1982).

The Estado Novo implanted some of the necessary infrastructure. In 1942, the Companhia Siderurgica Nacional (National Steel Company) built the first large-scale iron- and steel-producing plant in Brazil. The plant was constructed in Volta Redonda in the Paraíba

River Valley with North American financing, obtained in return for Brazil's support for the Allies in the Second World War. In the same year, the government created the first public enterprise to mine iron ores: the Companhia Vale do Rio Doce (Rio Doce Valley Company, CVRD). This was also financed by the United States and its Ex-Im Bank. With the creation of the CVRD, the Brazilian government became the owner of ore deposits which had belonged to the English group called Itabira Iron Company.

The role of widening the country's productive base – whether by establishing enterprises in basic industry to break bottlenecks in energy, transportation, and mineral extraction, or as a regulator of the labor market through complex labor legislation – fell to the state. The process of industrialization, however, continued to be constrained by the nation's import capacity.

Economic, political, and ideological disorganization gradually placed economic policy and planning in the hands of the government, strengthening its interventionism, without threatening the power of the dominant groups. Thus it presented a conservative pattern of modernization which was to characterize the country's future development.

The affirmation of the authoritarian ideology under Vargas' dictatorship made state-building more relevant than ever before in Brazilian history. National will was placed above class interests, justifying the need for a strong government to prevent social fragmentation (Skidmore 1973). The authoritarian ideology provided justification for both strengthening stateness and politically incorporating new social sectors to enhance nationhood (Reis 1983). In terms of state-building, the expansion of the military and bureaucratic apparatus during this period clearly emphasized the significant process of power growth and centralization. The Armed Forces were modernized, their personnel increased, and the military became actively involved in policy. The creation of the National Security Council, composed of the executive power and the Armed Forces, was the most obvious expression of the assumption that national security required economic growth. This would become, in time, a "natural" premise in Brazilian politics.

The state guaranteed a power coalition which accommodated traditional agrarian elites and the emergent industrial sector. The power of the former derived mainly from the continuing social relations of production in the countryside which kept rural labor outside the political arena. In turn, the industrial elites benefited fundamentally from the state's economic policies. Finally, the state ensured its own position by politically incorporating the popular urban sectors. The state opened a political avenue for urban workers, especially through social rights, guaranteed by an extensive and modern labor code, even though labor unions were controlled by the Ministry of Labor and social and political stability was assured by severe repression.

The unity of the national territory became a chief symbolic resource for state legitimation. And the territorial policy, incorporating and accelerating dynamic social and spatial tendencies, became one of the foundations for state practice. Once again, the availability of land supported authoritarianism, but now in a new context: industrialization based on low salaries made possible by a dynamic frontier. At the same time, the Vargas government started a campaign for a "Westward March," that is, the conquest of Brazil's "empty spaces" comprising all the interior located "behind" the coast.

National-developmentalism (1945–1967)

In the period following the Second World War, corresponding to phase A of the fourth Kondratieff cycle, the world-economy registered a vigorous growth attaining planetary dimensions under the new hegemony of the United States. The US dollar became the international currency, allowing multinational corporations to expand throughout the world. Brazil was directly affected by this process for two reasons. First, its position in the immediate area of influence of the United States; second, a deliberate state policy to internalize the dynamics of the world-economy in favor of a national development project.

The external constraints to national development

The United States became Latin America's only source of capital and technical and military assistance, as well as being Latin America's most important market. Almost 60 percent of Latin American imports between 1946 and 1948 originated in the United States, which also absorbed almost half of all the continent's exports. But, in spite of the North American monopoly, its foreign policies did not correspond to the Latin American elites' expectations of hemispheric cooperation.

The highest level of expectations with regard to this cooperation occurred in Brazil after Vargas fell from power in 1945, by a military coup which, paradoxically, assured the promulgation of a liberal Constitution in 1946. Brazil was the most loyal US ally; its collaboration in the war effort had gone well beyond the cession of air and naval bases within its territory; it also assured the supply of basic primary materials and strategic materials at stable prices during the conflict; and Brazilian forces were involved directly in expelling German troops from Italian territory in 1944. Brazil was one of the first countries in the world to participate in discussions concerning the creation of international institutions projected by the United States for the post-war world. In 1945–46, then, the majority of Brazil's civil and military elites considered themselves a privileged partner (within the Latin American region) in the construction of the new international order which was outlined under the economic, military, and political hegemony of North American capitalism. They considered themselves a special case, meriting financial help from the United States in promoting Brazil's development (Malan 1986).

Nevertheless, the "special relationship" was essentially asymmetrical and generated enormous frustrations in Brazil (as in the rest of Latin America). The United States' real priorities were in Europe and Asia, aimed at containing Soviet expansion. In line with its project of internationalizing the world-economy, the United States insisted on the global, and not regional, character of its post-war policies. Faced with financial requests to promote Brazilian development, the North American position counseled Brazil to fight inflation, eliminate restrictions on international trade, and encourage private firms (especially foreign) by creating a political climate appropriate for international flows of private capital – a debate which continues even today.

The return of Vargas, by election in 1950, suggested a new national-populist period. The creation of Brazil's National Bank for Economic Development (BNDE) in 1952, as a result of the suggestions of the Brazilian–United States Mixed Commission, did not stop the Vargas government from imposing restrictions on remitting profits and the return of capital, in an attempt to stabilize the trade balance and to affirm its nationalist posture, thus provoking a break with the World Bank. This break imposed serious restrictions on the Brazilian economy's development, aggravated by the steep decline in coffee prices after 1954.

It became clear that industrialization by import-substitution, dependent on the export capacity of the agricultural sector, tended toward stagnation. In this context of crisis, ECLA's thesis became relevant: since the market forces were insufficient to guarantee the development process of Latin American peripheral countries, state planning was an essential path for national industrialization. This thesis was coincident with the aims of the military interested in the question of security and development, as well as of a small part of the Brazilian elite interested in diversifying the productive structure and in accelerating the pace of capital formation in industry and in productive infrastructure. They believed that structurally transforming the Brazilian economy would require a significant degree of state intervention in economic life as well as a significant contribution of foreign technology and resources – both official and private.

The substantive issues of economic policy which continue today were almost all related to disagreements over the *form* and *extent* of foreign and public-sector participation in economic life. Conflicting development projects emerged. The first one defended a form of national capitalism with state intervention, illustrated by the creation of the state petroleum company, PETROBRAS, in 1953. This was a victory for a wide range of social sectors, and for the National Bank for Economic Development. The other approach sought an association with foreign capital to liberate industry from its dependence on the agro-exporting sector. The latter project was ultimately victorious.

Associated development: the triple alliance

The redefinition of world capitalism and the new orientation of internal social forces converged in implanting a new model of accumulation. In consolidating Brazilian capitalism, however, it redefined and deepened the country's economic dependence. "Fifty Years in Five" and "Energy and Transport" were slogans of the Kubitschek government (1956–60), revealing the importance given to expanding the nation's investment capacity, to the factor of accelerated time, and to the physical means of production. An industrialization policy favorable to private monopolistic capital was adopted – a politically directed capitalism called *desenvolvimentismo* (developmentalism).

The new model represented a rupture with the prior economic orientation on two levels (Mendonça 1986). The first level was the redefinition of the industrial sector, privileged by the state which favored durable over non-durable goods. Imports were limited by an unfavorable balance of payments, so the economy could not face the demands of the increased internal market associated with urbanization and its new consumption patterns.

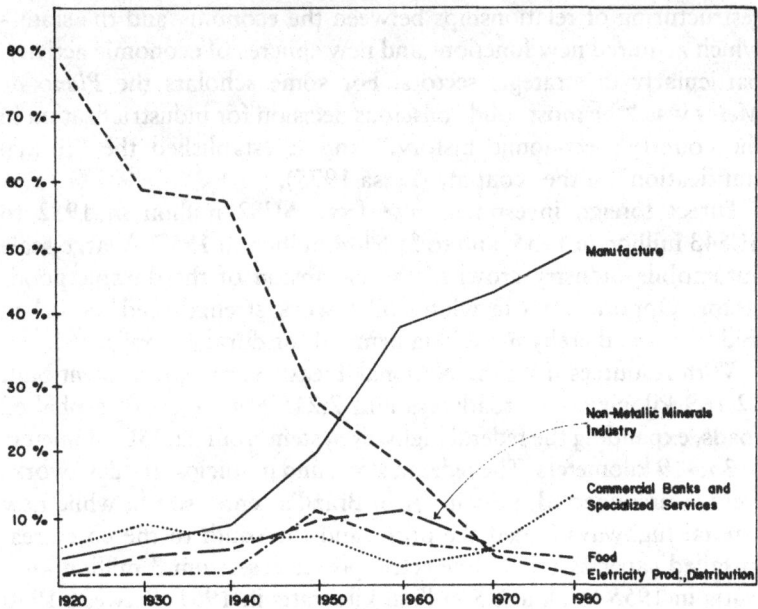

Figure 2.6 Foreign capital investments by economic sector, 1920–1980.
Adapted from *Retrato do Brasil* 1984.

The second level could be found in new financial patterns. The choice was made to internationalize the Brazilian economy, through direct investment. Once the US directed its investments toward European industry, the Brazilian government adopted a policy of attracting the capital necessary for overcoming the balance of payments crisis and for implementing the model of associated development.

European recuperation, due to massive US investment in its manufacturing industries, led to alterations in the international division of labor which were vital for industrialization in Brazil, as in other peripheral countries. After 1957, direct foreign investment and supplier's credits for importing machinery and equipment became easier because of competition between European and North American suppliers. The roots of the *tripe* model were then established: national private capital produced non-durable goods, foreign capital dominated production of durable goods, and state capital operated in the sphere of production goods (Figure 2.6).

This association was controlled by the state through planning. The *Plano de Metas* (1956–60) was a landmark in the qualitative restructuring of relationships between the economy and the state – which acquired new functions and new spheres of economic activity, particularly in strategic sectors. For some scholars the *Plano de Metas* was "the most solid conscious decision for industrialization in the country's economic history," and it established the "formal statification" of the economy (Lessa 1975).

Direct foreign investment rose from $US2 million in 1952 to $US43 million in 1955, and to $US144 million in 1957. A large-scale automobile industry crowned the expansion of the durable goods sector. Opportunities in white-collar work strengthened the urban middle class, thereby increasing demand for durable goods.

With resources from the National Treasury, the government built 12,169 kilometers of roadways and 7,215 kilometers of asphalted roads, expanding the federal highway system from 22,250 kilometers to 35,419 kilometers. The federal, state, and municipal road networks were interconnected, principally in Brazil's center south, while new arterial highways joined the north and the south to the core area. Installed capacity for hydroelectric power rose from 3 million kilowatts in 1955 to almost 5 million kilowatts in 1961. Between 1950 and 1960, oil production jumped from 2 million barrels/year to 30 million, and steel production grew from 1,150 tons annually to 2,500. There were thirty state-owned enterprises in 1949, but forty-five in 1960. In 1960, 321,150 vehicles were produced in the country by a few large foreign-owned plants. Modern industry surpassed agriculture (Figure 2.7). A whole cycle of import-substitution was accomplished, a process that represented an outstanding example of "closed" late industrialization (Serra 1982), even though it occurred in association with foreign capital.

The new model had many negative outcomes as well. Concentration of capital and firms, inflation and external debt, income concentration and increasing bureaucratic power, were all problems that arose during the Kubitschek administration, but exploded only afterwards. Economic crisis broke out in 1962, and was exacerbated by the political crisis which came upon the resignation of the populist President Janio Quadros (elected in 1960). In 1964, the military took power in a coup, but they were unable to bring about economic recuperation until 1968.

Political legitimacy during the Kubitschek government was attained through party alliances, traditional bargaining methods, and some concrete concessions to the working class regarding salaries and political participation. However, the most active element in neutralizing social tensions was the ideology of national-

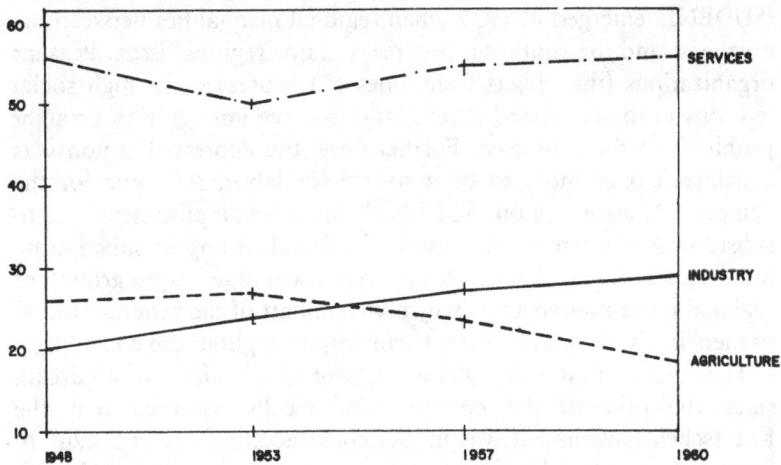

Figure 2.7 Brazil: gross domestic product by economic sectors, 1948–1960 (percentages).
Basic data from IBGE 1990a.

developmentalism, including territorial articulation. This ideology had a dualistic vision of Brazilian society – a traditional agrarian sector against a modern urban-industrial one – and elevated the national bourgeoisie to be the revolutionary vanguard for overcoming underdevelopment. Actually, the national entrepreneurs' own project was not a defense of autonomous industrialization but rather a gentlemen's agreement with foreign capital. The nationalistic discourse tried to engage all of society in the process of "national capitalism," hiding the contradictions that were going on beneath the opening of the economy.

Territory was both a tool and a product of "national capitalism," through implicit and explicit state spatial strategies. Regional planning emerged. Developmentalism strengthened urbanization's role as the basis for industrialization, again favoring economic concentration in the southeast. The priorities of federal policy for industry and infrastructure were developed by channeling resources from agriculture. Most direct government investment – in industry, transportation, and energy – went to the southeast, due to a compromise with interests in the southeastern states. Foreign corporations favored locations within and around the metropolitan area of São Paulo, while Rio de Janeiro – the locus of the bulk of state companies – experienced a pronounced decline.

The Superintendency for the Development of the Northeast

(SUDENE) emerged in 1959 when regional inequalities between the northeast and the southeast became a major regional issue. Peasant organizations (the "Ligas Camponesas") expressed the high social tensions in the depressed region, and massive immigration brought problems to the southeast. Furthermore, the depressed region was considered once more to be a source for labor, this time for the "empty" Amazon region. SUDENE's main accomplishment was to subsidize southeastern companies who installed import-substituting industries in the northeast. This process encouraged rapid growth of regional metropolitan areas which were a part of the general process of metropolitan concentration occurring throughout the country.

The construction of Brasília in the central *cerrados*, an old dream since the time of the empire, symbolically signified that the Kubitschek government was in fact constructing a "New Brazil" in five years, and had legitimate power over the entire national territory. It had a political significance, isolating the central power from pressures by the "coastal" society, particularly from the urban masses of Rio de Janeiro. It also had an economic significance. Situated in a central strategic position – in contact with the coast and the backlands and at the nexus of dynamic, stagnant, and unpeopled areas – the new capital became a center for the great highways giving access to the south, east and northeast, and the gateway for entering the north and the west.

In this process, the spatial structure of an "archipelago" gradually gave way to a "center–periphery" structure. Relations with the world-economy are fundamental to explaining the regional conformation of contemporary Brazilian spatial structure. But it is also a result of the regional policy, as an instrument for extending the state's political space over the national territory.

3

The world-economy and Brazil's regions

The aim of this chapter is to describe the spatial-temporal dynamics of the process of differentiation/incorporation of the Brazilian regions in the world-economy. Its starting point is the *colonial production space* where the Portuguese dominance, performed through the commercial monopoly, implanted mercantile settlements in the coastal privileged places. These establishments, directly subordinated to metropolitan control, were neither countryside nor cities, but the rural and urban face of the great ultramarine enterprise of the old colonial system.

The emergence of domestic commercial interests, linked to slave traffic and smuggling, was the origin of the mercantile cities that gradually acquired autonomy. Establishing a direct relationship with the Prata Basin, as well as with England, this rising mercantile bourgeoisie contributed to the breaking of the "colonial pact" and to the formation of the *mercantile archipelago* in the nineteenth century.

Each agro-exporting region belonging to the archipelago displayed within it a separation between mercantile city and agro-pastoral countryside. The city, as a bourgeois–commercial locus, acted as intermediary between the producing rural zone and the international circulation of products. Given the mercantile form of this insertion of the Brazilian economy, each region came to specialize in one or two export products, which made the territorial division of labor resemble the pattern of Brazilian exports.

Although dependent on the cyclical fluctuations of world trade, the mercantile regions displayed the interests of social classes identified with the territory. That is, they had recognized the source of surplus which permitted the valorization of invested capital. Thus it is possible to characterize these parts of Brazilian space as regions of the world-economy, subject to centralized national political control under the hegemony of the economic group which was then dominant: the coffee-producing bloc.

The industrialization process, intensified from 1930 on, broke off with the relative isolation of the mercantile archipelago articulating the Brazilian regions under the command of the São Paulo dynamic center. The internalization of *center and periphery* relations in Brazil was an unequivocal demonstration that the laws of motion of industrial capitalism acted within the national space. It also showed that the process of Brazil's incorporation into the world-economy was complete – since it was already reproducing internally the same mechanisms which were acting at the global level.

Brazil's spatial configuration at the beginning of the 1960s displayed an integrated but uneven structure. The largest share of national income was centered in the southeast, given the significant concentration of industrial production in that region. By contrast, Brazil's other regions were organized into a vast periphery in which each part carried out particular functions in the new territorial division of labor resulting from the country's industrialization.

The colonial production space

The process by which the Brazilian territory was incorporated into the world-economy was marked by a social and territorial division of labor into three large sectors: the *marinelands*, which corresponded to the lands dominated by slave agriculture near the shores; the *backlands*, a vast hinterland complementary to the coastal economy where extensive cattle-raising constituted the principal economic activity and in which, from early on, labor relations were not dominated by forms of slavery; and the *mines* which represented a dense settlement of particular parts of the interior and which sustained important commercial flows with the coast as well as with the backlands.

In this period, the presence of the metropolitan state was marked, not only in the consolidation of the territory, but also in the configuration of the colony's spatial structure. The cities and towns were points of tribute and control over the colonial economy. The city was not a place of "citizens," nor a village of "freemen," but rather a "locus" of power under the Portuguese kingdom. In this sense, it is not possible to speak of a separation between country and city, both being part of the same colonial establishment of mercantile capital.

The social division of labor was dominated from outside. The historic evolution of the socio-economic colonial formations is one which makes it possible to understand under which conditions the interests which gave the ancient colonial economies their own dynamic came to be crystallized.

The long history of the colonial period is also a history of the emergence of the first forms of local power, capable of imposing the marks of its control over the territory. The conditions for the formation of these structures of power, identified with the appropriation and valorization of surplus in a certain point of space, revealed not only the contradictions inherent in the peripheral form of insertion into the world-economy but also the conflicts between diverse political and social interests.

The marinelands

Knowledge of the Brazilian coast was the first effective result of the colonizers' presence. This undertaking was realized through various official expeditions as well as those financed by the Portuguese and Dutch commercial bourgeoisie. Within a mercantile framework, they sought information about the wealth available on the Brazilian coast, in particular the existence of exotic products with high unit value in the European markets and metals and precious stones. Subordinated to this overriding interest, they also sought to discover new routes to Asia and to locate points along the coast which could serve as supports for boats traveling on the Route to the Indies. This was particularly important because of the wind patterns south of the Equator which, given the technical limitations of overseas navigation in the sixteenth century, could lead to the loss of an entire fleet imprisoned in the calms of the Atlantic.

The natural means of communication privileged the eastern coast from the start, in terms of the objective of economic exploitation. This was accentuated by the presence of an immense humid forest in this part of the Brazilian coast, the Tropical Atlantic Forest, formed by the precipitation coming from the moisture carried by the onshore winds (Figure 3.1). The forest was the natural "habitat" for various indigenous tribes of the Tupi-Guarani group, who had left a cultural phase of hunting and gathering and had situated themselves at the beginning of an agricultural revolution, involving the domestication of various tropical plants – such as manioc, corn, and tobacco, among others – which were cultivated in small open fields within the forest (Ribeiro 1977).

The colony's first economic activity was focused on the forest and the labor of the indigenous communities: extraction of dyewoods, especially brazilwood (*pau-brasil*). Collected from the forest by the Indians and stored in *feitorias* – fortified waystations along the eastern shore, brazilwood inaugurated the institution of the *estanco* – the King's monopoly over commerce with the colony – and also the constant visits by French and English corsairs.

The marinelands were the territory of the *engenho*. The first efforts to implant a sugar economy in Brazil tried to utilize the basic conditions set by the extraction of dyewoods. First, the tropical forest furnished wood for construction and fuel for the mills. Second, the moist coastal plain guaranteed the river transport of sugar production in addition to offering soils which renewed their fertility with the periodic rains; and finally an attempt, which was rapidly frustrated, was made to use the indigenous communities as sources of compulsory labor to open land and service the plantations.

The great slave plantation, which was the backbone of the colonial economy, was always restricted to the coast – whether because of the costs of transport, which made production at long distances from the ports non-viable, or because of its essential dependence on the fertile soils of the tropical forest. Based on this productive system, the great slave plantation was established at various points along the Atlantic coast. Meanwhile, due to diverse factors which were not merely geographical but also political and economic, this system reached its apogee in four areas of the Brazilian coastline: the northeastern "Zona da Mata" – centered on Olinda and later on Recife; the "Reconcavo da Bahia," around Salvador; the rim of Guanabara Bay, centered on Rio de Janeiro; and finally, the Gulf of Maranhão, on which São Luís was founded.

Salvador, the current capital of the State of Bahia, well illustrates this process. Situated in one of the best-protected anchorages of the eastern coastline and surrounded by a belt of fertile land, this city from early on became a preferred "locus" for establishing mercantile sugar enterprises. Headquarters to the Governor General of the colony until the middle of the eighteenth century and a large center for slave trafficking, Salvador was also responsible for the emergence of one of the first activities originating in the colony itself – the cultivation of tobacco for exchange with slaves from the African coast.

The role of Olinda, situated in the eastern northeast, was very similar, and perhaps more important given its closer proximity to the

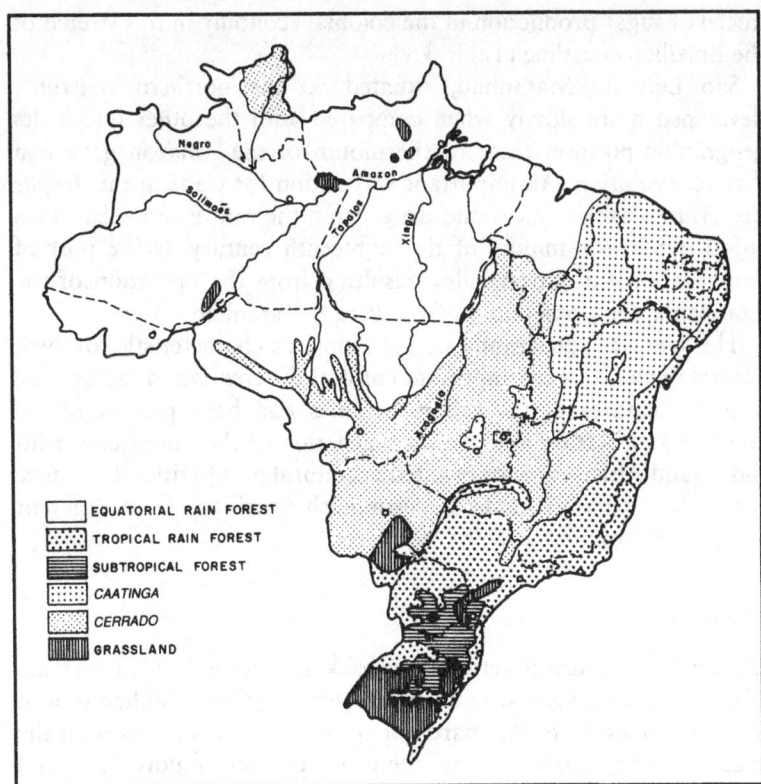

Figure 3.1 Brazil: native vegetation.
Adapted from IBGE 1972.

centers of consumption in Europe. Headquarters of the Capitania of Pernambuco, one of the most prosperous of the colony, its wealth attracted the Dutch who founded Mauricia, currently Recife, within sight of Olinda. The Dutch remained there for a long period during the Iberian Unification (1580–1640). The rivalry between Olinda and Recife was also the scene of the first native movements of the colony: the War of the Mascates – the name given to the merchants of the port of Recife whose interests were opposed to the large landowners residing in the hills of Olinda.

Rio de Janeiro was founded in an extremely favorable site from the point of view of mercantile enterprises, Guanabara Bay. It was initially of interest to the French, who in the middle of the sixteenth century created Antarctic France. They were quickly expelled by the Portuguese, however, who developed one of the most important nuclei of sugar production in the colonial economy in this stretch of the Brazilian coastline (Table 3.1).

São Luís do Maranhão, situated on the northern coastline, developed more slowly when compared with the other nuclei. Its geographic position close to the mouth of the Amazon gave it a distinct evolution. An important waystation for trade in the *drogas do sertão* – exotic spices and drugs from the rain forest – São Luís stood out in the middle of the eighteenth century as the port of embarkation for cotton bales, resulting from the operation of the Companhia de Comercio do Grao Para e Maranhão.

The historical, geographical, and economic characteristics of these coastal nuclei were an important cultural differentiation factor. The peculiar features of the urban localities had been present in the colonial architecture and the miscegenation of the Portuguese with Indian and African groups resulted in cultural peculiarities that, more than ethnic traits or dialects, distinguish Brazilians from different areas of the country (Ribeiro 1977).

The backlands

During the seventeenth century, the backlands were the scene of wars of extermination against remnant indigenous tribes who had moved inward, fleeing from the march of colonists. On its limits with the sugar-growing coastal zone there were also important social struggles. The *quilombos* – free territories of blacks seeking refuge from slave labor – were alternatives allowing liberty for the slave population. The most celebrated of these, Palmares, located in the Serra da Barriga in Alagoas, resisted various attempts to destroy it by the colonial authorities. It was finally annihilated by troops from São Paulo.

When the Portuguese established themselves in São Vicente – a territory corresponding to the current State of São Paulo along the middle of its coast – and began to cultivate sugarcane with enslaved indigenous peoples, the *bandeirantes* (pioneers) specialized in capturing Indians for the plantations. This practice stimulated exploration into the vast highlands which begin beyond the narrow coastal strip and the escarpment of the Serra do Mar. In the highlands, an extensive river network oriented toward the interior permitted the pioneers from São Vicente to spread out in all directions, capturing Indians, searching for gold and precious stones, and establishing cattle ranches.

The backlands were the territory of the curral. Cattle-raising,

Table 3.1. *Sugar production in Brazil, 1710*

Producing regions	Sugar mills	%	Production (crates)	%
Bahia	146	28	14,500	40
Pernambuco	246	47	12,300	33
Rio de Janeiro	136	25	10,220	27
Total	528	100	37,020	100

Source: Antonil 1963.

which was introduced to the colony in the neighborhood of the plantations, was crucial to the functioning of mercantile enterprise. Furnishing animals for draught, leather, and meat, cattle-raising in Brazil began simultaneously with the sugar mill. Meanwhile, the growing land requirements of sugarcane cultivation were gradually pushing the corrals farther into the interior, since the cattle could arrive at the colonial markets by their own motive power. Gradually driven away from the coast, husbandry found in the Brazilian interior two natural pasturages which allowed it to expand vigorously following the end of the seventeenth century: the northeastern *caatinga* and the southern grasslands.

With Salvador and Olinda as the generating nuclei, northeastern cattle-raising covered the semi-arid backlands (*caatingas*) through *ribeiras*, that is, ranches established along the rivers in an extensive ranching system which was founded upon share-labor. In this system, a rancher received a certain number of livestock from those born under his supervision. Ranching stimulated salt extraction along the northern coastline for manufacturing *charque*, salted meat dried in the sun, whose production was destined for consumption by the coastal slave population. The salt and dried-meat producers were responsible for the occupation of the semi-arid coastal zone. The technology developed in this area was carried to the extreme south, which later during the mineral boom came to be the principal zone for dried-meat production in the colony.

Beyond its important historic role in the occupation of the territory and its support for the sugar complex, ranching carried out another role – no less relevant in the configuration of the northeastern region: the consolidation of the latifundios which had been initiated by sugar. Endowed with its own characteristic of "natural" accumulation and a high power of resistance to the multi-century crisis of the export sector, ranching contributed greatly to the formation of an immense reserve of labor in the northeastern back-lands. This resistance derived, in large part, from its capacity to enclose itself in subsistence activities, creating virtual "mini-systems," expanding herds of livestock and the working population by mere natural growth. This autarchic process in an area of semi-arid climate, subject to periodic droughts, aided the occurrence of virtual "migratory waves," expelled from the northeastern backlands by droughts (Furtado 1959).

In the south, cattle-raising served as the economic basis for the occupation of the gaucho plains, where there were abundant lands and good pasture. Two fronts of ranching pioneers established themselves in what is now Rio Grande do Sul. The first was formed by Spanish Jesuits who introduced cattle and horses in the missions of Tape and Uruguay, in the early seventeenth century. The missions were attacked by *bandeirantes* from São Vicente aiming to capture Indian slaves as well as draught animals. These actions opened the way for the second front of Luso-Brazilians, in search of the vast extensions of southern fields for cattle-raising, in order to supply the mines' activities.

The mines

The discovery of gold in the highlands of Minas Gerais at the beginning of the eighteenth century started an intense mining cycle which, in contrast to the sugar plantations, took place some hundreds of kilometers inland and promoted the multiplication of small towns near the ore-rich streambeds and mines. Here, masters, slaves, officials, soldiers, and clerics depended directly upon the import trade for their subsistence, since the topography and the exclusive preoccupation with lucrative mining permitted neither space nor time for any other activity.

In contrast to sugar production, which was only accessible to those who were able to mobilize voluminous financial resources, panning for gold could be undertaken at the artisanal level as well as in large enterprises. The Portuguese immigration toward the region occurred on a much larger scale than had taken place in the two preceding centuries. Urban life, therefore, developed and created a market for foodstuffs, which came to reinforce the already more important market for draught animals destined for the extensive system of transportation connecting the vast gold-producing region to the port of Rio de Janeiro, the most important commercial center of the colony (Table 3.2). This market for livestock was principally

Table 3.2. *Rio de Janeiro: trade in commodities*

Market	Imported by Rio	Exported from Rio
Buenos Aires, Rio Grande do Sul	Hides Tallow Dried meat Silver Wheat	Sugar Cotton cloth Slaves Tobacco Rice Manioc
Bahia/Pernambuco	Coins Coconuts	Manioc Beans Cane brandy Hides Corn Minas cheese Wheat Dried meat Bacon Cotton Rice
Portugal	Wine Olive oil Brandy Vinegar Onions	Sugar Indigo Rice Cane brandy Hides Coffee Fish oil Wax Woods Cotton Tallow
Europe (other than Portugal)	Woolens Cutlery Clothing Prints (cotton) Dairy products Hardware Housewares Porter	
Africa	Wax Oil Sulphur Woods Slaves Ivory Salt	Sugar Rice Dried meat Tobacco Cane brandy Manioc Gunpowder Arms Cotton Bacon
North America	Salted meat Wheat Furniture Pitch Tar	
Asia	Silk Luxury goods	Silver

Source: Johnson 1972.

supplied by the southern regions, whose ranching capacities were already well known. In this way, the mining pole encouraged the formation of economic links between the northeast, the center, and the south of the Brazilian territory, as early as the eighteenth century – that is, in the period immediately preceding independence (Furtado 1959).

During the colonial period, then, mineral extraction shaped the beginnings of integration among the producing areas since – in order to increase control over gold extraction and to avoid contraband – the Portuguese metropolis transferred the colonial government's headquarters to Rio de Janeiro (1763). It thus centralized access to the mining area and, with this, Rio de Janeiro assumed an extremely privileged position in colonial commerce, and contraband. In the same way, the inhabitants of the town of São Paulo, founded on the highlands, benefited directly from the mining activities by becoming the principal source of draught animals, principally mules, brought from the south and distributed to the mines (Figure 3.2).

The decline of mining coincided with a crisis in the colonial pact. Its impact was revealed in the important role, as much political as economic, which Rio de Janeiro assumed – first as the seat of the Portuguese Court during the Napoleonic Wars, and immediately afterwards as the capital of the empire of Brazil. Due to its commercial intermediation for the mining areas, Rio de Janeiro already had capital seeking valorization independent of agricultural property, that is, it represented the embryo of Brazilian mercantile capital.

The mercantile archipelago

Brazil came to be inserted into the new international division of labor which resulted from the Industrial Revolution in the form of a mercantile-slave empire. It was linked to the capitalist world-economy through the circulation of products, since the productive process was controlled by mercantile capital and its labor relations were predominantly those of slavery.

In this period, the spatial structure of Brazil looked like an archipelago of mercantile regions, virtually drainage basins, with centers in the large port cities. The prime example was the mercantile-slave region of Rio de Janeiro, the principal coffee-producing area until the last quarter of the nineteenth century. The slave matrix of the coffee economy limited internal trade between the mercantile regions, whose direct links were instead oriented toward

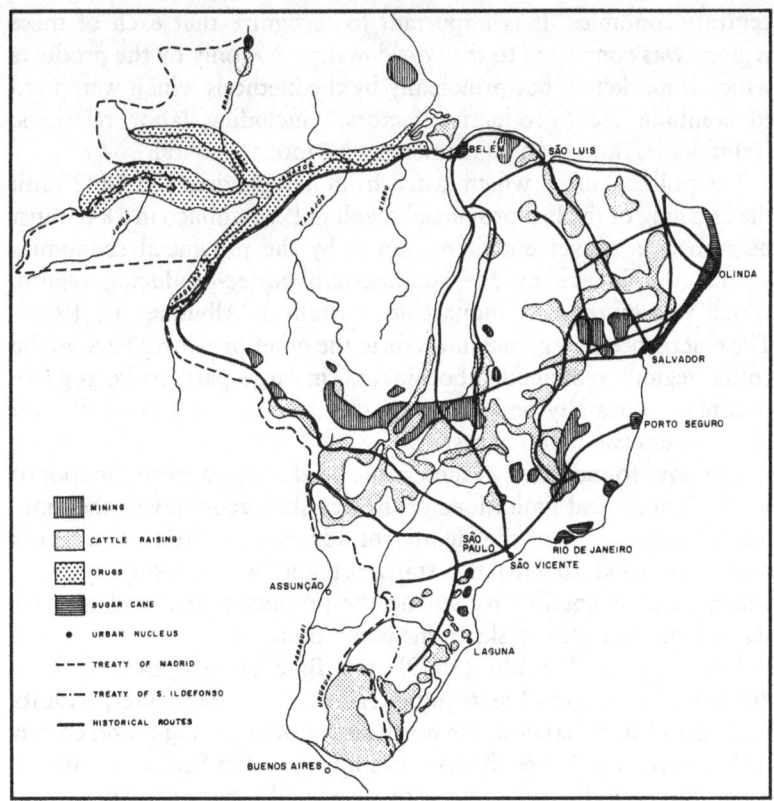

Figure 3.2 The colonial production space.
Adapted from Albuquerque, Reis, and Delgado de Carvalho 1980.

the world market. One should note, however, some internal links through the internal slave trade, which persisted until abolition, and supplies for the grand cities which, even though consuming largely imported products, did constitute a potential internal market which was gradually occupied by domestic agriculture and manufacturing.

The formation of an archipelago reflected Brazil's insertion as a producer of commodities for the world market – whether sugar, tobacco, cocoa, rubber, or coffee – utilizing the natural and historic comparative advantage of each part of the national space. Even though coffee was by far the dominant product, it is important to emphasize that all of the regions were parts of the world-economy because they articulated their connections with the world market through interests established here, although determined by the central economies. It is important to recognize that each of these regions was connected to the world market not only by the products which it marketed, but principally by the methods which were used to combine the "productive factors," including labor relations, technologies, and even adjustments with foreign capital.

The political crisis which lasted from independence in 1822 until the crushing of the last provincial revolt in Pernambuco in 1848 must be seen as a movement of resistance by the provincial economies against dominance by the southeastern coffee-producing region, which was favored by monarchic centralism (Albuquerque 1981). The emergence of regional interests in the other provinces beyond the coffee regions remained subordinated, in large part, to Portuguese commerce, formally dependent on the structures inherited directly from the ancient colonial system.

The slave foundations of mercantile production were at the root of the basic technical limitations of the Brazilian economy in the nineteenth century. All of the significant advances in productive forces were channeled toward the transport and warehousing systems, which made it possible to expand the producing area and transfer slaves from transport tasks to the plantations.

According to Furtado (1959), the Brazilian empire could be divided into five grand mercantile regions: the coffee center, with its nucleus in Rio de Janeiro; the northeast, producing sugar and cotton and centered on Recife; Bahia – headquartered in Salvador – which began the period as a producer of sugar and tobacco but gradually transformed itself into a cocoa exporter at the end of the nineteenth century; the south, oriented toward ranching and the processing of dried meat; and finally, the Amazon, which assumed growing importance in Brazil's external commerce in the final quarter of the nineteenth century through the export of natural rubber, and which was centered in Belém and secondarily in Manaus (Table 3.3).

The northeast and Bahia

The process of political separation between Brazil and Portugal was carried out by a particular fraction of the bloc of hegemonic classes which represented the interests of the coffee center. The divergence between the northeastern groups, involved in the historical crisis of the sugar economy, and the "Coffee Barons," in ascension in Rio de Janeiro, manifested itself right at the beginning of the empire with the Equatorial Confederation (1824), a republican separatist movement involving the provinces of the northeast from Alagoas to Ceará. This

Table 3.3. *Economic and demographic indicators of Brazilian regions, 1841–1850*

Region	% of total population	Demographic growth rate (%)	Income growth rate (%)
Northeast	35	1.2	−0.6
Bahia	13	1.5	0.0
South	9	3.0	1.0
Center	40	2.2	2.3
Amazon	3	2.6	6.2
Brazil	100	2.0	1.5

Source: Furtado 1959.

movement must be viewed as an attempt to restrict the influence of the centralizing bureaucratic apparatus and to resist the commercial and financial dominance of the Portuguese commercial bourgeoisie (Albuquerque 1981).

The Civil War in the United States – which interrupted the flow of North American cotton to Europe – permitted booms in cotton production in the Brazilian northeast. This expansion of cotton production made possible the development of an economic alternative to sugar – whose spatial mobility was severely restricted to the coastal band. Cotton, by contrast, demanded little from the soil and less in terms of capital investment. Cultivated with free labor and inducing an important branch of industrial processing, it seems that cotton, kept in due proportion, was a more dynamic economic activity than sugar, in terms of urbanization, commerce, and industrialization. Since the principal profits from cotton came from processing and foreign commerce, these became the preferred areas for mercantile capital which came to control the processes of cleaning and baling in the northeastern backlands. Exports were concentrated in the ports of São Luís, Recife, and Fortaleza, defining urban basins which captured the cotton processed in small units dispersed throughout the backlands. The relevance of Recife as head of a vast mercantile region was reinforced by the expansion of the railroads (Figure 3.3).

As the chronic crisis of the northeastern economy was accentuated during the nineteenth century, the expansion of coffee cultivation in the center transformed the northeast into an exporter of slave labor for the coffee plantations. In consequence, the abolition of slavery was anticipated in this region. Meanwhile, instead of implanting a capitalist labor market, the region developed forms of servile labor with payments for the use of land being worked, the *morador de condição* (Andrade 1973). Far from expanding the consumer market, this labor form led the regional economy to stagnate at close to subsistence levels.

Bahia's political role can be appraised when one recognizes that the Coat of Arms of the empire of Brazil was adorned with a branch of coffee and another of tobacco – both symbols of the agricultural "vocation" of this region south of the Equator. Bahia had been, since the colonial period, the principal tobacco-producing area in Brazil. Planted in large and medium properties around the bay, tobacco required – in fundamental contrast to sugar – regularly fertilized soils. From the start, this required that it be cultivated in association with livestock-raising. The decline of sugar activities allowed tobacco cultivation to expand and, in 1877, tobacco took first place among Bahian exports, creating with it an important processing industry in the areas around Salvador.

Cocoa, planted in the south of the province where climatic conditions were similar to those of the Amazon, developed slowly during the entire nineteenth century, becoming important at the beginning of the twentieth century. Initially a modest crop developed along with subsistence activities, its evolution toward monoculture followed the process of land concentration unleashed by the commercial bourgeoisie centered in Ilhéus. This process configured the most explicit form of mercantilism in Brazilian territory where labor, required to open the plantations and harvest the crops, stagnated in small towns dispersed through the cocoa plantations.

The center and the south

It was the capital accumulated in commercial intermediation with the mining zone which financed the emergence of the economic activity which became the trademark of the empire of Brazil: the slaveholding coffee plantation. The Brazilian coffee boom emerged from a combination of several contributing factors. Commercial capital accumulated in Rio de Janeiro was able to finance producers through the five years required before the trees begin to bear fruit. Also, slave labor became available with the decline of mining; and the forested lands available in the neighboring areas of Rio de Janeiro were excellent for growing coffee.

Beginning with plantations in the massive coastal areas of Guanabara Bay, coffee rapidly took the Paraíba Valley, a tectonic *graben* between the Serras do Mar and the Serras da Mantiqueira.

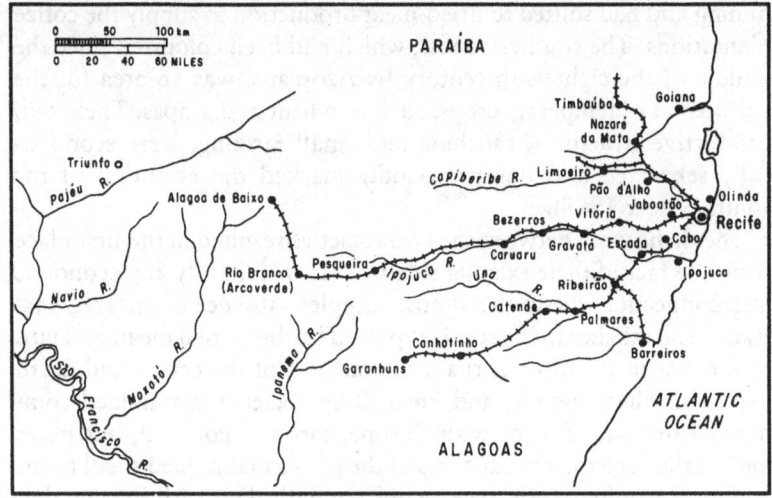

Figure 3.3 Recife's mercantile region.
Source: Levine 1971.

The valley is a virtual corridor between Rio de Janeiro and São Paulo and was well traveled by the mule trains which supplied the mines. The railroad was an alternative which made it possible to reduce transportation costs in the coffee region, as well as in other mercantile regions, increasing the mobility and volume of products for export. Meanwhile, in no other Brazilian producing zone did the railroad have the same impact as it had for the coffee economy: linking the plantations of Rio de Janeiro, opening new plantations, allowing the transfer of slaves from transportation to agricultural work, and bringing modernization to the interior (Figure 3.4).

If on the one hand the railroad consolidated the spatial links between the plantations and the ports of Rio de Janeiro and Santos, shaping a vast area which extended from the forested mining areas to the São Paulo highlands, on the other hand it revealed the gradual failure of slavery in Brazil. The railroad, which represented the implantation of a productive force originated and developed under wage-labor relations of production, showed itself to be incompatible with slave-labor. The speed of the trains accentuated the crisis of slavery and opened the path for wage-labor, as did the coffee-processing machinery which was gradually incorporated into the plantations as a way of substituting for the "lack of hands for cultivation," according to the discourse of the period.

In the south, ranching had suffered directly from the decline of mining and had shifted to dried-meat production to supply the coffee plantations. The southern coast, which had been colonized since the middle of the eighteenth century by Azoreans, was an area for the cultivation of temperate crops, such as wheat and grapes. These two productive structures, ranching and small farming, were economically separated and this profoundly marked the evolution of the south (Singer 1968).

The dichotomy between the two societies resulted in the first place from the fact of their existing side by side with hardly any economic interconnection due to the almost complete absence of intraregional trade. The ranches in the south exported leather, dried meat, and lard to the rest of Brazil or abroad; the farmers of the center and north exported wheat, liquor, and linen. Both societies maintained commercial ties with Rio or with Europe, through Portugal, but never among themselves. The ranchers of the plains maintained small farms and artisanry for subsistence, acquiring with their cash income that which they could not produce in their own domains. The Azorean farmers and later the German and Italian colonists behaved similarly. Nothing indicates the existence of economic integration in Rio Grande do Sul at least until the current century, when industry began to take shape.

The Amazon

Beginning in the second half of the nineteenth century, rubber extraction emerged as the principal activity of the equatorial forest. The seringueira (*Hevea brasiliensis*) is a tree which is native to the region and which, given the great diversity of species in the Amazonian environment, is dispersed throughout the interior of the forest. This obliges the rubbertappers who collect the latex to make long daily trips to obtain a profitable amount of natural rubber.

The activity of extracting latex expanded slowly in the 1860s and 1870s, attaining a more significant level in the 1880s and reaching a peak between 1890 and 1920. The increase in international prices, which tripled between 1880 and 1910 as a consequence of the utilization of natural rubber in manufacturing automobile tires, stimulated the expansion of rubber extraction. The rapid expansion was possible because of the entrance of immigrants from the northeast who fled the droughts and the restrictions on the labor market in their places of origin. This migratory movement was particularly intense after the great drought of 1877, which struck the semi-arid interior of the northeast.

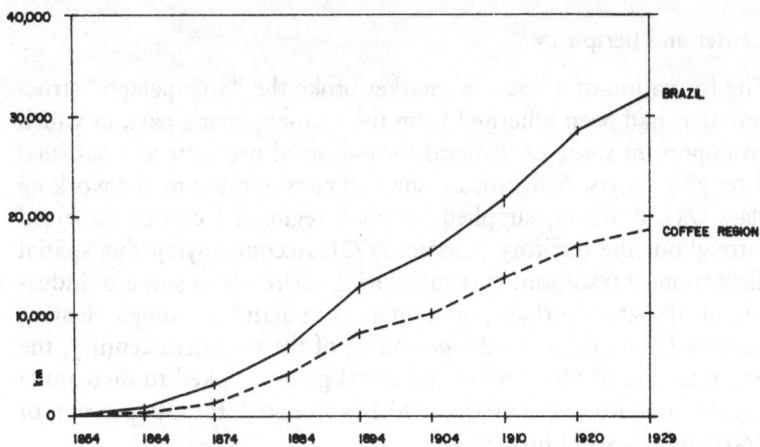

Figure 3.4 Railroad expansion in Brazil, 1854–1929.
Basic data from Silva 1976.

Given the conditions peculiar to the technique of extracting and processing rubber, as well as its orientation toward the external market, the Amazonian capitals functioned as centers for a great drainage system connecting the rubbertrees to foreign markets through a vast river network. In Belém and Manaus, large exporting firms were established which monopolized the rubber market. These firms were subsidiaries of German, English, North American, and French companies who dominated the export market and, gradually, came to dominate the diverse internal agents involved in production.

The peak of the rubber boom was directly reflected in these two cities since they were virtually "islands" in the middle of the forest, and the surplus resulting from the exports was transformed into sumptuous buildings and into consumption of all kinds of imported luxury goods. Outside these cities little happened, since no significant investments were made – neither in the transportation system which demanded only small ports and docks for the construction and repair of river boats, nor in the very cultivation of regional foodcrops because the majority of consumer goods in the cities of Belém and Manaus were imported and the rubbertappers practiced subsistence agriculture and fishing to feed their families. The collapse of the rubber economy was precipitated by the competition of the southeast Asian plantations, dropping the region back to a primitive economy.

Center and periphery

The formation of a national market broke the "archipelago" structure that had been inherited from the agro-exporting past, in which an important share of demand for industrial products was satisfied through imports. A significant share of consumption by the working class was, however, supplied by small regional factories dispersed throughout the territory (Castro 1971). Accompanying this spatial dispersion of traditional manufacturing, there was a surge of industries in the grand urban port centers with activities complementing imported commerce. At the beginning of the twentieth century, the best example of this kind of industrial growth linked to mercantile capital was Rio de Janeiro, which accounted for 30 percent of Brazilian industrial production.

This situation began to change, in the sense of breaking regional isolation and initiating a process of regional articulation between the diverse producing areas of the national space, with the expansion of coffee cultivation based on wage-labor in São Paulo. In contrast to previous situations, the purchasing power of Paulista exports was not completely transferred abroad through imports, as occurred at other times and in other regions. Rather the demand created by coffee cultivation opened the possibility of interregional exchange of merchandise, since the needs of the working class – a majority of whom were colonist immigrants – for food, clothing, and footwear, were partially satisfied by industries located in diverse parts of the national territory.

Sugar from the northeast, lard and wine from the south, and textiles from Rio de Janeiro were among those products which encountered a market in the wake of the expansion of coffee, gradually conquering the distances between the economic "islands" through modernization and expansion of the transportation network laid down by the mercantile empire. The dynamism of the coffee economy also generated a complex interregional network. Investments in infrastructure aimed at making the expansion of coffee production viable also created external economies which benefited the formation of an industrial park, rather diversified and with relative autonomy in relation to mercantile agro-exporting commerce.

The consolidation of an industrial nucleus within the coffee complex introduced significant modifications to the Brazilian spatial dynamic. Among these, the change in the role of the agricultural frontier assumed special importance, since until then it had responded primarily to external impulses. With the beginning of industrialization, the dynamic vector of territorial expansion came, in large part, to serve the needs of the industrial center. The particular conditions of industrial development in Brazil, as analyzed in Chapter 2, favored the spatial concentration in the coffee region, despite several dispersed initiatives in some coastal cities.

Changes in the spatial organization of Brazil's economy accompanied the substantial modifications in the way Brazil was inserted into the world-economy. Brazil participated in the formidable expansion of the world capitalist system after the Second World War, no longer simply as an exporter of commodities but – given the marked presence of the state in supplying infrastructure – as a location for investments by national and multinational firms.

The consumer-durable industry, which had been the motor of industrial growth during the 1950s and 1960s, required a market of national dimensions. This was only attained, however, with the imposition of protectionist trade barriers against imported products and the conquest of regional markets, previously supplied by importing from abroad. The ability to claim a share of the world market for Brazil's nascent industry was a demonstration of the power of the state apparatus, which came to represent the interests of the industrial bourgeoisie. The federal government eliminated the privileges of regional groups, through negotiation, thereby forming a wide-ranging alliance legitimated by national-developmentalism.

The massive bloc of investments in the *Plano de Metas* acted upon a spatial structure inherited from the past, and it resulted in a double process. On the one hand, it accentuated the historic concentration of economic activity in the southeast by completing the territory's vertical integration of industry through fixed capital investments in the generation and distribution of energy and in basic industrial inputs. On the other hand, it modernized and amplified the peripheral network, expanding the dimensions of the internal market.

The role of the periphery was fundamental in sustaining the pace of growth for industrial activities, principally in that it continued to limit the wage share of industrial production costs. The continuous supply of labor guaranteed by increasing migratory flows toward the southeast, along with the elastic supply of food given the growing incorporation of land through expansion of the agricultural frontier, assured industrial growth without great inflationary pressures. The role of the periphery in the new territorial division of labor was driven internally by the very process of capital accumulation, since maintaining low wages was a key element in guaranteeing increasing profit rates in industry.

In this phase, the expansion of the frontier appears to be linked to the need for increasing agricultural production with a low level of capitalization, so as not to impede urban-industrial capital accumulation. The increase in agricultural production was, therefore, attained through extensive rather than intensive expansion under a primitive structural form of accumulation in which the surplus created by the transitory occupation of the land by rural workers or small producers was appropriated and transferred to the dynamic center.

Cattle-raising played a fundamental role in agrarian organization oriented toward the internal market, as it came to replace the cultivation of export crops in organizing new spaces and in reorganizing old ones. It offered important advantages by opening the possibility of appropriating large tracts of land with few workers, allowing the production of food to form pastures, reducing transport costs by herding cattle to market, and resisting inflation. With the new mode of accumulation, then, a new territorial division of labor began within agriculture, with a tendency to dissociate food production from crops destined for export. In the face of the relative limitations on capital, the internal market, and the transportation network, and with the characteristics of the existing spatial structure as well, this division of labor took its clearest and sharpest form in the dynamic region.

The regional structure resulting from industrialization could be characterized by three units: the core area with the integrated periphery, the depressed peripheries, and the resource frontier. These units represented not only geographically differentiated regions, but also distinct historical moments, when all forms of production were present – from simple production marked by profound interdependence on natural resources, to complex urban-industrial structures which were the fruits of a rapid accumulation of social labor (Figure 3.5) (Becker 1982).

The core area and the integrated periphery

In 1960, the core area and the integrated periphery represented metropolitan Brazil, generating close to 85 percent of national income and intense flows of goods, labor, and capital. Differentiation within it already indicated tendencies of regional specialization in the form of an industrial society. Within this region one can

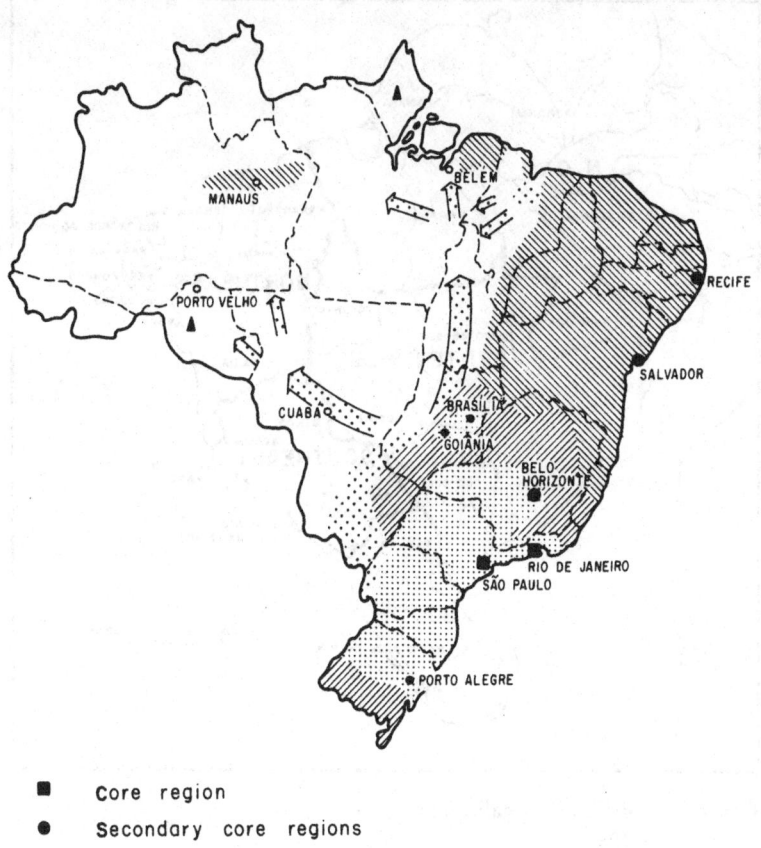

- ■ Core region
- ● Secondary core regions
- • Emerging core regions
- Upward transitional or developing regions
- Slowly developing regions
- Downward transitional or depressed regions
- Frontier resource or new opportunities regions
- Pioneer fringe over woodlands near the core region
- Main directions of the pioneer fringe
- ▲ Mineral exploration

Figure 3.5 Types of region according to spatial interactions in the 1960s.
Source: Becker 1982.

Figure 3.6 Official regionalization.
Source: IBGE 1967.

distinguish the core area, the dynamic, and the slowly developing peripheries.

The core was the polarizing nucleus of the southeast – a region which emerged from coffee growth and industrialization from the ancient center and was defined in the official regionalization in the late 1960s (Figure 3.6) – corresponding to the national metropolises, São Paulo and Rio de Janeiro. The area was characterized by diverse forms of organization of urban life and industrial activities. Among these were the large industrial plants, such as in the Paulista metropolis; industrial nuclei, as in Volta Redonda and Cubatao; and a string of smaller industrial centers tending to establish themselves along the Paraíba Valley, as well as in the direction of the Paulista interior, actually forming a developing region. It was surrounded by a dynamic agrarian belt on the western part of the Paulista highlands,

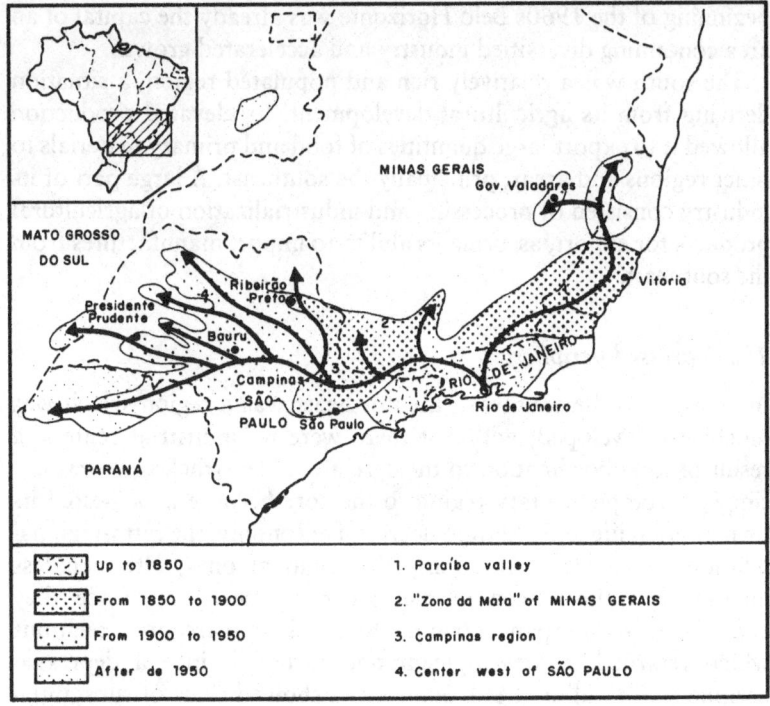

Figure 3.7 The march of coffee.
Source: Rodrigues 1977.

aided by the presence of wooded lands, favorable climatic conditions, rolling topography, and rich and fertile soils in particular stretches. It included as well the northern highlands of Paraná, which became the principal area for coffee in the 1960s (Fig. 3.7).

In contrast to the polarization of the São Paulo region, the area under Rio de Janeiro's influence presented a traditional agrarian economy where activities were not diversified after the decline of the coffee plantations. Large ranches and small farms predominated there, producing milk and milk products for supplying the metropolis. Sugarcane was cultivated in large properties in the north of the State of Rio de Janeiro, which with the *Zona da Mata* of Minas Gerais and the State of Espírito Santo formed a slowly developing region near the core area.

Still within the southeast was the area of mining and metallurgy in the center of Minas Gerais, where large steel mills were built. They were soon followed by other kinds of industries, so that by the beginning of the 1960s Belo Horizonte was already the capital of an area containing diversified industry and accelerated growth.

The south was a relatively rich and populated region, a situation deriving from its agricultural development. Its elevated production allowed it to export large quantities of food and primary materials to other regions and areas, principally the southeast. A large part of its industry consisted of processing and industrialization of agricultural products for export, assuring its ability to import manufactures from the southeast.

The depressed peripheries

In contrast to the southeast, the northeast was a region which was much less developed; within it there were no industrial centers, a result of its subordination to the core area. Nevertheless, it was not simply a complementary region to the core because it possessed its own internal life and a certain degree of autonomy: the intra-regional relations were larger than the inter-regional ones, although less intense than those seen in the integrated periphery, since they concerned an agrarian economy founded in mercantile relations which retarded its development and turned it into a depressed periphery (Carvalho 1988). Some cities showed signs of substantial manufacturing activities, Recife being the most important, although they consisted almost exclusively of food-processing and textile plants.

The polarizing nuclei of the northeast were located within the coastal band, with complementary regions in the middle north, the west of Bahia, and the north of Minas Gerais. Nevertheless, the absence of a solid and diversified industrial base kept the northeast from connecting more solidly with these areas – suffering from growing competition with the southeast, which tended to polarize the country as a whole.

The unfavorable physical conditions in a large part of the north-eastern territory – due to irregular rainfall – restricted agricultural expansion, limiting it to the humid coastal band. The agrarian structure displayed archaic forms of organization, and productivity in general was much lower than in the dynamic periphery. In 1960, industrial production in the region comprised only 8 percent of Brazilian industrial output, and was largely of non-durable consumer goods. Artisanry still played an important role in this part of the economy. The demand for more complex industrial products was satisfied in large part through imports from the southeast, whose influence in the interior of the northeast never ceased to grow (Oliveira 1977).

The strong social tension – associated with a long drought period – that marked the end of the 1950s led to a direct intervention of the state through SUDENE. SUDENE's basic proposal was the integration of the northeast region into the dynamic core by opening opportunities of investment to southeastern industrial capital through tax and credit incentives. The SUDENE experience was later generalized to the Amazon and to the center west. Regional planning became an instrument for the state to speed up the mobility of the financial capital in the Brazilian territory incorporating new areas into the dynamic core and through it into the world-economy.

Traditional local societies in the Amazon and in the extreme south ranching zone also constituted depressed peripheries, due to the permanency of ancient social and economic structures, in which the source of wealth and power lay in landownership and control over the commercial links. The depressed-periphery dominant groups played an important role in the maintenance of the power bloc and in the resistance to the "basic reforms," mainly agrarian, that was the claim of the social movements of the early sixties.

The resource frontier

The introduction of cattle in the center west can be linked to the open vegetation of the savannah, since the wooded lands were originally destined for agriculture. The grand domination of the savannah suffered the reorganization of its traditional activities through the formation of artificial pasture, improved herds, and the introduction of wire fencing for separating fields. The region's proximity to the southeast was vital for its occupation and for the entrance of farms into the woodlands. Cereals became an important regional product, particularly with the introduction of rice in Goiás which complemented the supply of foodstuffs for the urban-industrial centers.

The north was above all a largely untouched region, not an "empty space," but one sparsely populated by Indians, rubbertappers and a traditional local society. It did not have any nucleus capable of commanding a more complex regional organization. The extractive activities in the forest were economic activities of greatest importance. Internal relations were very tenuous, practically reduced to the transportation of products for export and the distribution of imported merchandise. From this, the cities of Belém and Manaus

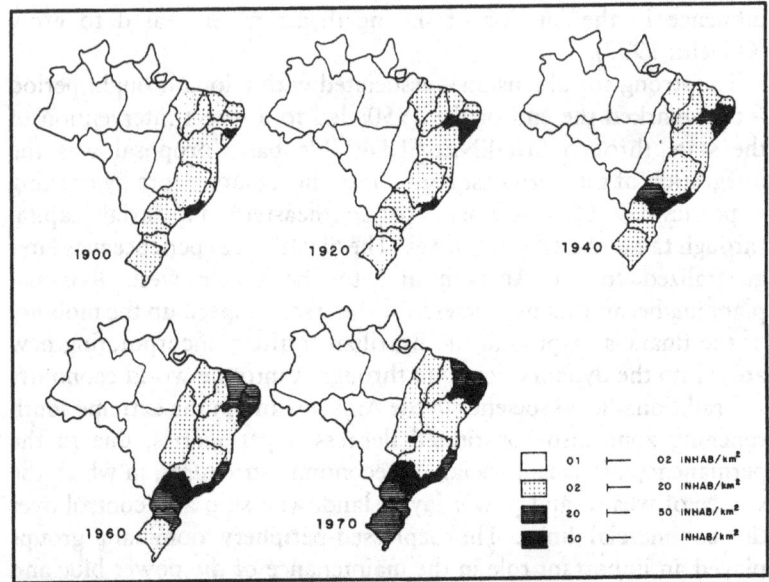

Figure 3.8 Demographic densities in Brazil, 1900–1970.
Basic data from IBGE, demographic census.

developed disproportionately; oversized, they dominated the regional space.

A special agency was created in 1953 to promote economic activities and build an infrastructure in the Amazon. Its main accomplishment was to construct a pioneer road from Goiás to Belém – the future Belém–Brasília highway – which expanded the agricultural frontier into the eastern Amazon, and attracted immigrants from all parts of the country, seeking land and jobs. It also stimulated intense land speculation.

The construction of Brasília and the Belém–Brasília highway marked the opening of the north frontier to the dynamic national center in the southeast (Valverde and Dias 1967). Cattle-raising spread in the north of Goiás, accelerating the expansion of the resource frontier which, in addition to the discovery and commercial extraction of mineral resources – such as the exploitation of manganese in the Territory of Amapá and cassiterite in the Territory of Rondônia – created urban nuclei (Katzman 1977) and economic enclaves within the vast forest, which remained, in large part, sparsely occupied (Figure 3.8).

Brasília symbolized the eagerness for integration of a nation which, huddled for centuries on the coastal strip, has gazed back into the immense unpopulated interior which was gaining value in the new economic and social conjuncture. The new capital, situated in a strategic position, in contact with every type of periphery, represented a true spearpoint of the "center." It stimulated both the advance of the pioneer fringe and the economic links with São Paulo.

4

The rise of Brazil as a regional power in the world-economy

In the 1970s, Brazil altered its position in the structure of the world-economy and passed into the category of semiperiphery as a regional power. This process began during phase B of the fourth Kondratieff wave when two interrelated processes came to propel the transformation of the world capitalist system: the radical crisis/restructuring of its patterns of accumulation, until now based on Fordism and Taylorism, and the technological revolution, principally in micro-electronics and information science. Science and technology reformulated the bases of power which came to emanate from accelerated speed, or rather, from control over space and above all over time (Virilio 1977).

Multinational corporations and large transnational banks impose a new international division of labor in which nation-states are ceasing to be the economic units of the new historical reality. But they remain as political unities and states condition the restructuring of the economy for which the continuous preparation of the means of war assumes a significant role (Castells 1985).

Notwithstanding the global recession, multinational industrial decentralization, and abundant credits offered by large banks – combined with domestic specific conditions – produced a deep differentiation on the peripheral sector of the world-economy. Brazil, Mexico, the Asian New Industrialized Countries (NICs), China, and India experienced a short cycle of growth between 1967 and 1982 rising in the world-system as semiperipheries, a process sustained by external indebtedness and by a vigorous state intervention.

The change in Brazil's position was attained thanks to preexisting conditions such as its large territory, the significant internal market, and a solid industrial base laid down in the earlier phase. But it was also the fruit of deliberate policy enacted by an authoritarian regime which was socially exclusive. The military conquered the state, which took upon itself the execution of a geopolitical project toward modernity.

The new authoritarianism was a strategy to accelerate development of Brazilian capitalism, but also a strategy of the state for the state itself. The combination of the geopolitical project with the historic authoritarianism resulted in a conservative modernization implying profound transformations and contradictions which ended by destabilizing the regime, at the beginning of the 1980s. It is this conservative modernization, which is immediately responsible for the country's emergence as a regional power, that will be analyzed in this chapter. Its legacy will be treated in the next chapter.

The geopolitical project toward modernity

The premises of the geopolitical project were not determined by the country's geography nor were they merely aimed at physically appropriating the territory. The mark of the new project was the intention to steer the modern scientific–technological vector toward controlling time and space. This control was understood by the Armed Forces as a condition for the construction of the nation-state in this new world era, and for the accelerated modernization of society and national space which was necessary to attain economic growth and international prominence. It was also a condition for consolidating and expanding the leading role of the state – under the tutelage of the Armed Forces – understood as the only actor capable of accelerating modernization through rational planning (Becker 1988). The political role and relative autonomy of the Armed Forces changed qualitatively. They ceased to be a "bureaucracy in arms" and became instead "the armed designers and managers" of a national scientific–technological project.

This is not meant to credit the military with omniscience. Initiated under the liberal regime of the post-war period, the geopolitical project was already implicit within the *Plano de Metas* of Kubitschek's government; and it was not the fruit of the Armed Forces alone, but rather of several fractions of the civilian and military elite. It was not a rational and intelligent campaign, but rather a collection of isolated initiatives and decisions taken according to the conditions at hand, full of dilemmas, which ended by converging upon a governmental project managed by the military.

However, two aspects of their actions need to be stressed: (a) the acknowledgment that the capital-goods industry alone was not sufficient to guarantee national sovereignty, which also requires technological autonomy; (b) the instrumentalization of space as a basis for accumulation and state legitimation.

The conservative modernization combined these two aspects and treated space as an integral and fundamental part of the technical base of the tripod model, seeking to endow it with operationality and functionality capable of guaranteeing not only the expanded repro- duction of the diverse interests involved, but also the integration of portions of the national territory as privileged areas of valorization within the world-economy (Egler 1988). In its turn, the territorial integration was an essential ideological resource utilized to expand the control over the national territory and to hide selective social and spatial policies.

The state transformed previous historic conditions, producing its own space as regulator and organizer of national territory. It seeks to control economic stocks and flows by imposing a spatial order connected to the idea of global space which is rational, logistic, and of general interest (Lefebvre 1978). This idea – a representation constructed by the state techno-structure – contradicts the practices and conceptions of the local space of private interests and the particular goals of diverse social agents. It creates, then, a global– fragmented space; global because it is technically homogenized – facilitating the interaction of spaces and moments – but fragmented because it is appropriated in pieces, as argued theoretically by Henri Lefebvre (1974, 1978).

Geopolitics became an explicit doctrine which was at the same time a justification for and an instrument of the state's strategy and practice. In consonance with the goals of the project, the govern- ment's strategy concentrated its force upon three time-spaces with specific practices: (1) the implantation of the scientific–technological frontier in the country's core area; (2) the rapid integration of the entire national territory, implying the definitive incorporation of the Amazon; (3) the projection of the nation into international space.

The National Security Doctrine and the new authoritarianism

Taking power in 1964 and retaining it for twenty-one years, the military carried out their project systematically, thereby abandoning its earlier "moderating" policy and initiating an authoritarian period of a new kind. In this period, the industrial and financial bourgeoisie joined together and obtained legitimacy among the middle class, excluding some of the most backward sectors of the dominant groups from the power bloc and squashing the popular sectors.

There is a great difference between the traditional authoritarianism exercised by the generals and dictators who are still present in some Latin American societies and the new authoritarianism which emerged in the societies which had modernized by the 1960s. In the new authoritarianism the military dominated as an institution – not at the level of the individual – and with a techno-bureaucratic approach to formulating policy. From this comes its designation as bureaucratic-authoritarianism, a situation in which the Armed Forces took power to restructure society and the state intervened against popular movements maintaining the continuation of "progress" and "development" according to the modern military doctrine of National Security. This situation was not characteristic of all the military regimes of Latin America, but only of those of the southern cone – Brazil, Argentina, Uruguay, and Chile – in which the decisive factor was the militarization of the state (Cardoso 1979).

There is no consensus over the reasons for the emergence of the new authoritarianism. One explanation is predominantly economic. In this view, repressive governments were a response to the difficulty of "deepening" the process of industrialization – that is, of developing the production of intermediate inputs and capital goods. They therefore promoted accelerated industrial growth through the further concentration of income and by imposing a form of social and political development which was "exclusive and concentrating" (O'Donnell 1973).

In the case of Brazil, however, which could be considered a paradigm for the authoritarian military model, the "deepening" of industrialization had already been effected, and by democratic governments. In Brazil, on the one hand, political-ideological factors were a key influence on the implantation of the new authoritarianism. The dynamic of class struggles in the early 1960s, marked by strong popular demands and the proposal for basic reforms, the climate of the Cold War, the Cuban Revolution (1958), the fear of the dissemination of guerrilla-war tactics and the determination of the United States to impede a second Cuba, all these created the perception of a heightened degree of threat among the leading groups of the military institution (Serra 1979; Cardoso 1979; Hirschman 1979). On the other hand, organizational and timing factors also weighed heavily. Once again, the political role of the Armed Forces can be explained in large part by their capacity for organization in contrast to the ideological debility and indecisiveness of the political behavior of civilian groups and classes, and the weakness of popular organizations in a developing society (Carvalho 1988). Organized and pursuing a project, the military in this way conquered the state.

As an institutional expression and instrument of the new project, the Escola Superior de Guerra (ESG) was established in 1949 with the support of the United States, using the standards of the American National War College and the French Institut des Hautes Etudes de la Défense Nationale (Hepple 1986: 28). The new military perspective materialized in the formulation and practical implementation of the National Security Doctrine, a strategic planning technique initially designed for use in the field of national security policy in times of war, but which was to be extended to all sectors of activity in the country (Figure 4.1).

The concept of National Security is the kernel of the doctrine. According to this doctrine, the fight for survival requires that the country maximize economic growth. To attain this growth, security is necessary and it also imposes sacrifices on the people. The doctrine is stated most explicitly in the works of General Golbery do Couto e Silva (1955, 1957). It involves the national perspective of an underdeveloped country aiming at accelerating its development and reaching a new status according to the current model of advanced capitalist nations, under the tutelage of the state, but at the same time maintaining a vision of relative autonomy within the hemisphere in the face of US hegemony.

For many, the influence of the ESG doctrine in the 1964 coup and in the military government which followed was central, displaying a "new professionalism" (Stepan 1973). For others, the distinctive element of the 1964 coup was not ideology but growing military and technical capacity to control and repress (Markoff and Baretta 1985). Still others suggest that this was not a fully instituted authoritarian "regime," but rather an authoritarian "situation" constructed according to the available formulas of legitimation and permeated with dilemmas and fissures (Linz 1973; Lamounier 1988). Nevertheless, the doctrine gave a degree of legitimacy and intellectual and political structure to the military.

Seeking technological self-sufficiency

According to Hirschman (1986), the policy of "import-preemption" industrialization refers to the decision to bypass the import-substitution sequence and encourage the domestic manufacture of

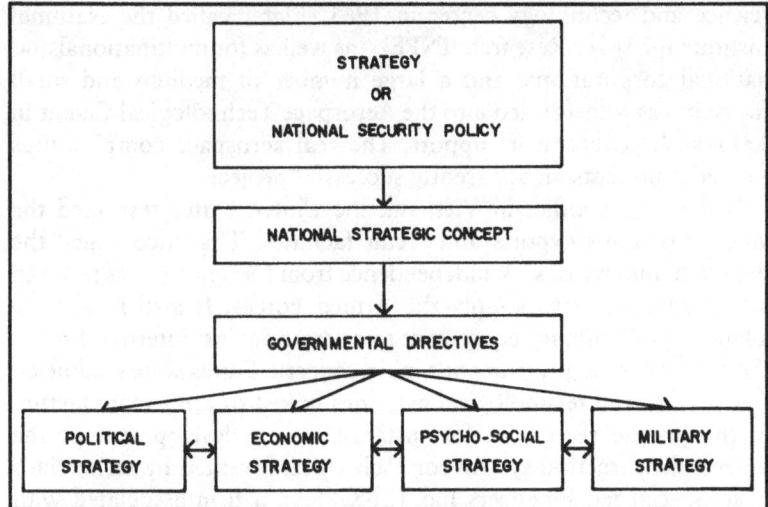

Figure 4.1 The National Security Doctrine as a strategy.
Source: Couto e Silva 1981.

new products. This occurred when the Armed Forces aimed at dominating the modern scientific–technological vector, particularly in four strategic sectors: aircraft, weapons, nuclear and computer industries.

The Brazilian aircraft industry is the result of a process initiated with the creation of the Air Force Technical Institute (ITA) in 1946, which later in 1951 became the Air Force Technology Center (CTA), with the explicit purpose of linking teaching, research, and industry. The technological autonomy of an underdeveloped country was pursued thereby through a nationalistic ideology, which formed the basis for an articulation of the state with private corporations and the scientific–technological system (Dagnino 1983). The CTA has been the chief institution in Brazil for the formation of scientific–technological human resources – more than three thousand engineers – contributing as well to the development of other segments of both civilian and military industries.

In 1969, the government transferred the projects developed in the CTA to EMBRAER. The purchasing power of the government was essential, through an unprecedented coordinated action in national industry: the orders received by EMBRAER and other companies, mainly from the Air Force Ministry, sustained a minimal although stable level of production. The CTA was the attraction pole for a new science and technology center in 1961 – later called the National Institute for Space Research (INPE) – as well as for multinational and national corporations, and a large number of medium and small firms; it was transformed into the Aerospace Technological Center in 1971 with government support. The real aerospace complex thus formed represents an apparently successful project.

Due to the conflict in Vietnam, the United States restricted the amount of arms exports and credit facilities. This encouraged the Brazilian military to seek independence from foreign sources through an internal effort to supply the Armed Forces. It also meant the adaptation of military equipment to operate against internal threats, the authoritarian government's main concern. Funds were channeled to scientific and technological activities linked to arms manufacture in the private sector, as for instance in the development of the independent traction system for each wheel invented by technicians from Specialized Engineers Inc. (ENGESA), a firm associated with the Army. Protectionist measures and special incentives made the sector's expansion feasible.

The Army took its own route to modernity. Originally fragmented, it tried to bring rationality to its project through the creation of IMBEL in 1975, a state military corporation that attempts to produce the various kinds of equipment needed for ground forces through association with private companies. It stimulated the production of explosives as well as light armored vehicles and tanks. As a result, Brazil became an important weapons exporter (Table 4.1).

Nuclear research has its roots in the creation of the National Research Council (CNPq) in 1951, which was conceived by an admiral with the intention of developing a domestic nuclear program. Aiming at an atomic policy independent from the United States, a Brazilian–German agreement was reached in 1975. A state holding company – NUCLEBRAS – was given responsibility for the agreement's implementation. But this initiative was not successful, given the inadequacy of the enrichment technology (jet-centrifuging) and mistakes in the construction and operation of the project. The Navy's action since 1979 has been linked to the development of a silent "parallel" nuclear program in collaboration with the National Council for Nuclear Energy (CNEN) and the University of São Paulo (USP), a program different from the one that is continuing under the Brazilian–German agreement.

The Navy also played an important role in the national policy for computer and semiconductors industry. Since 1971, a working group

Table 4.1. *Shares in weapons' market, 1971–1985 (%)*

1971–75		1976–80		1980–85	
United States	39.2	USSR	39.7	USSR	34.0
USSR	36.2	United States	32.7	United States	25.2
Britain	8.5	France	11.4	France	13.9
France	7.8	Britain	5.1	Britain	5.3
China	2.4	Italy	2.4	Italy	5.0
Germany	1.2	China	1.7	Germany	4.1
Italy	1.1	Germany	1.1	China	3.5
Netherlands	0.7	Israel	1.0	Spain	1.6
Canada	0.5	Netherlands	0.9	Israel	1.2
Sweden	0.3	Brazil	0.8	Brazil	1.1
Total	97.9		96.8		94.9

Source: Le Monde Diplomatique, March 1988.

associated with the National Bank for Economic Development (BNDE) has aimed at the creation of a nationally produced computer. At that time, thanks to a strong domestic market and fairly advanced industrial engineering, the government decided to encourage the domestic computer industry. The first step was taken in 1977, when the government refused to give permission to IBM and other transnational corporations to manufacture micro- and mini-computers in Brazil.

In 1979, the Special Secretariat for the Computer Industry (SEI) was created, a special agency directly subordinated to the National Security Council in Brasília. It was charged with designing a national policy for the sector, temporarily reserving the national market for domestic firms – a policy which was reaffirmed in 1984. The micro-computer industry has experienced rapid growth and declining costs since that time. It has developed indigenous capacity for innovation and has become an important source of employment (Erber 1985; Evans 1985; Schwartzman 1985). It also developed into a conflictual issue with the United States.

It was in the Second National Development Plan (1975–79), in the midst of the first oil shock, that the import-preemption technological policy became clear. Due to the fact that associated research between universities and firms was not a complete success, science–technology efforts and expenditures were partly channeled to state and military research and development centers which assumed a central position in the structure of Brazil's scientific–technological development. State-owned enterprises multiplied or modernized to control strategic sectors such as mineral exploration, nuclear energy,

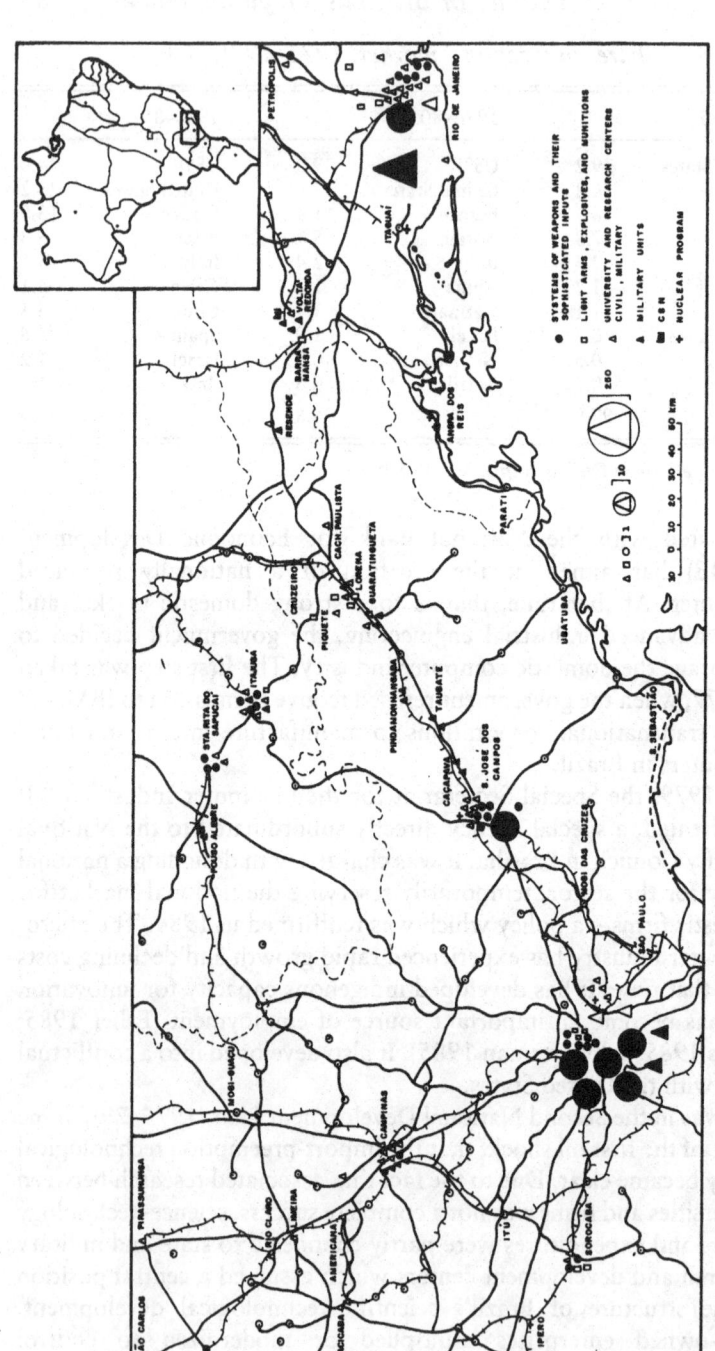

Figure 4.2 Concentration of the weapons industry in Brazil, 1990.
Source: Becker and Egler 1989.

telecommunications (with the creation of TELEBRAS), aerospace, oceanography, petrochemicals, electronics, and weapons. They are associated with research carried out in some private companies and/or university centers mainly in the State of São Paulo (Campinas, São Carlos, and the City of São Paulo) and in Rio de Janeiro (the Federal and Catholic Universities).

The "locus" of the modern geopolitical project is the South Paraíba Valley, the historical axis linking the metropolises of Rio de Janeiro and São Paulo, where the goal of building an industrial–military complex has become clear. The CTA was located in the valley, as well as ENGESA and the most important missiles manufacturer, AVIBRAS, a national private firm. Today, 80 percent of Brazil's weapons industries are concentrated in this valley (Figure 4.2).

The decision to locate the new project in this area may be explained by the valley's strategic position – both in economic and military terms – and by local favorable conditions in terms of land and the technical environment. The valley is the main route into the Brazilian plateau, and the metropolitan corridor through which flows the country's vital circulation. The valley's proximity to the military decision centers in Rio de Janeiro and to the industrial core in São Paulo is a primary factor of its strategic value. The technical personnel who graduate from military institutions which have existed there since the beginning of the century, as well as technicians drafted out of the large steel mill (CSN) created in 1942, were also essential locational factors. As a result, modern war and associated industries expanded throughout the valley and its surroundings, where the territorial division of labor is being redefined with science and technology as its basis.

Thus a new technological frontier emerged in the country's core area, bound to the new industrial production and its associated research and development centers. And this frontier was a necessary, although not sufficient, condition for the radical and accelerated restructuring of the territory.

In spite of these developments, government expenditure for science and technology represents hardly 1 percent of GNP according to the official budget, although the amount of research and development which is funded through the military ministries and state enterprises is unknown. On the other hand, technological dependence has not relaxed (Figure 4.3). This is due in large part to the "control" of the sector and the consequent lack of coordination between the technological policy and the economic policy.

Conservative modernization

The authoritarian military regime encompassed a very complex period. It included the crisis of the 1960s, the "miracle" between 1968 and 1972 and the onset of the recession which characterized the first years of the 1980s. In pursuing accelerated modernization, the state sustained high levels of investment with large government spending and direct intervention in the economy's productive apparatus, at the cost of indebtedness to the national and international banking system. Its program for modernization was also based on a territorial project informed by ideas of national integration and Brazil as a power.

The "Brazilian Miracle" and the economy in "Forced March"

From a strictly economic perspective, the coup of 1964 did not herald the definition of a new model of accumulation. The authoritarian modernization rested on the compression of wages and on the expansion of multinational, national, and state capital – consolidating the "Triple Alliance" (Evans 1979). But the state expanded its political and economic role throughout the country. On the one hand, there was competition among the subsidiaries of foreign oligopolies and between them and national private firms; on the other hand, there were projects requiring joint and interdependent actions involving the provision of infrastructure, raw materials, and basic inputs. Therefore, a complex state machine was mounted.

The organization of the governmental apparatus was crucial to reinforcing the centralizing mechanisms of the state. It was characterized by a double-sided process of concentration–deconcentration (Martins 1985). The process of concentration refers to the administrative and fiscal reform which expanded the extractive capacity of the federal government, giving it its own base for accumulation. The deconcentration process, realized through a complex institutional apparatus, refers to the extension of business activities by the state as producer and investor through multiple government agencies and autonomous public enterprises.

The recuperation of the economy which resulted in the "Brazilian Miracle" had two more basic conditions. The first was wage compression and control over the labor market, stimulating turnover and movement in the labor force. A new wage and labor policy assured low wage levels which were only readjusted from year to year, resulting in greater exploitation and an increased supply of

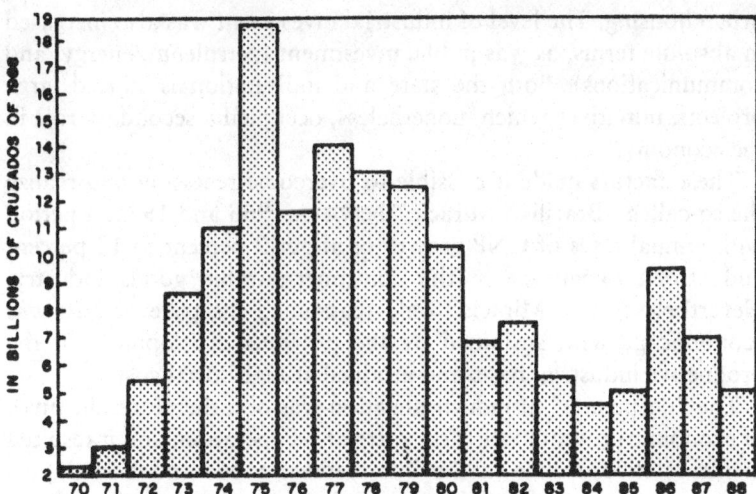

Figure 4.3 Research and development spending by the National Scientific and Technological Development Fund (FNDTC), 1970–1988.
Source: Klein and Delgado 1988.

labor: to maintain their living conditions, workers extended the working day (receiving paid extra hours) and intensified family labor. Labor turnover resulted in large part from social policy, which represented an exercise of control by the state over society in order to attain its economic objectives.

Mechanisms of compulsory savings, theoretically aimed at expanding the real wage, weakened employment stability in that they liberated the firms from any legal liability in firing employees, thereby resulting in high turnover and high spatial mobility. Furthermore, the extension of labor legislation to the countryside – the Rural Workers' Statute of 1963, and a land policy, the Land Statute of 1964 – resulted in an intense dislocation of workers and exploitation of peasants, associated with the concentration of land.

A second condition for the "Miracle" was the reinvigoration of the economy. Economic growth was still based upon the consumer-durable-goods sector, since that sector was already endowed with enormous potential for accumulation. The sector was also dominated by large firms which exercised considerable political pressure. Civil construction – which absorbs workers and does not require significant imports – was another element in economic reactivation. Its expansion was associated with the creation of a National Housing Bank (BNH) which channeled funds to the construction of urban housing. The level of industrial investment was also increased in absolute terms, as was public investment (petroleum, energy, and communications). Both the state and multinationals started large projects, initiatives which, nonetheless, occupied a secondary role in the economy.

These factors made it possible to overcome recession and realize the so-called "Brazilian Miracle" between 1968 and 1972: a period with annual rates of GNP growth of about 9 percent to 10 percent and strong expansion in the consumer-durable goods industry. Nevertheless, the "Miracle" only reassumed the pace of post-war economic growth, sustained by the previous development of the productive industrial base, and attained at high social cost.

The "deepening" of industrialization was not, therefore, the pivot of economic growth in the first years of recuperation, and it resulted in the relative decline in the intermediate-goods and basic-inputs sectors. In the face of rapidly growing imports, a crisis in the balance of payments emerged in 1973, which was aggravated the following year after the first oil shock.

It was in this context, when the country's needs for petroleum, raw materials, and machines revealed themselves to be far beyond its import capacity, that the new government of President Geisel (1974–79) established a program with the explicit goal of changing the dynamic core of the economy, from consumer durables to the sectors that represent the final stage of import-substitution industrialization: intermediate inputs and capital goods. In response to the crisis of 1973–74, instead of applying restrictive policies the economy was pressed into a long period of "Forced March" through the Second National Development Plan (II PND, 1975–79), which was initially seen in the maintenance of exceptionally high rates of investment in spite of the crisis (Castro and Souza 1985). It was possible to take this course only by borrowing, an option that became feasible in view of the large amounts of petro-dollars that became available and of vigorous "loan-pushing" carried out by the major banks in the 1970s (Hirschman 1986).

The II PND was the most important and concentrated state effort since the *Plano de Metas* of Kubitschek's administration to promote structural changes in the economy, an effort which took place at the exact time that the world-economy entered its severest recession since the 1930s. The program for projecting national power became explicit. The strategy for its achievement, inspired by the Japanese model, had as its central nucleus the strengthening of national firms; industrialization led by capital-goods production; growing

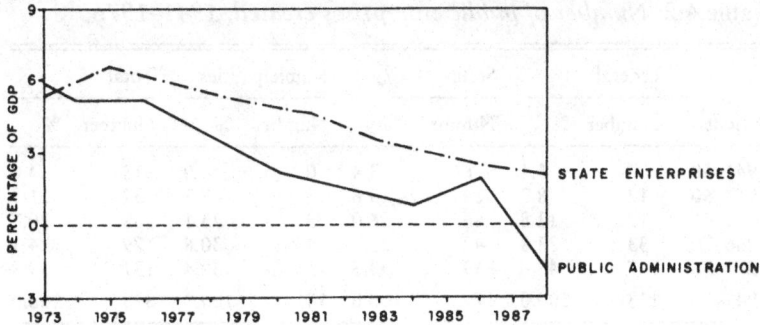

Figure 4.4 Public investments in Brazil, 1973–1988.
Source: Economic and Social Planning Institute (IPEA).

autonomy in the production of technology; encouraging financial conglomerates; and altering external relations to expand the degree of national economic independence, taking advantage of the conditions of the international crisis. The social policy, however, did not change in essence (Lessa 1979).

The program of import-substitution was accompanied by a vast energy program and the implantation of a petrochemical industry. It also included the effort to expand considerably the production of exportable raw materials (cellulose, iron, steel, and aluminium) by means of gigantic investments in the form of "joint ventures" between the state, multinational firms, and to a lesser extent private national capital.

The significant role of the state as an impulse factor for industrialization was exercised not only through its function as a provider of public goods, but also and above all (a) in defining, articulating, and financially supporting large blocks of investments which determined the principal modifications in the Brazilian economy's structure; (b) in creating an infrastructure which aimed at integrating the highway, energy, urban, and telecommunication systems; (c) in directly producing intermediate inputs which were indispensable for heavy industrialization (Figure 4.4).

The state enterprises were responsible for the decline in the imports of capital goods and equipment. In only six years, from 1970 to 1976, the number of federal companies doubled (Table 4.2), and by 1976 state enterprises accounted for 30 percent of the assets and 5.2 percent of the sales of the 5,300 largest non-financial firms. At the end of the 1970s, the public sector as a whole accounted for

Table 4.2. *Number of public enterprises created, 1941–1976*

Periods	Federal Number	%	States Number	%	Municipalities Number	%	Total Number	%
1941–50	7	5.1	6	3.4	0	0	13	4.0
1951–60	12	8.7	24	13.6	1	7.7	37	11.3
1961–65	19	13.8	46	26.0	3	23.1	68	20.7
1966–70	33	23.8	42	23.7	4	30.8	79	24.1
1971–76	67	48.6	59	33.3	5	38.4	131	39.9
1941–76	138	100.0	177	100.0	13	100.0	328	100.0

Source: Martins 1985.

almost 40 percent of the country's gross capital formation (Serra 1982).

The transnational corporations (TNCs) were concentrated predominantly in manufacturing, where they owned more than 30 percent of the capital stock. They controlled the most dynamic sectors, dominating the production of consumer durables (85 percent of sales), having a majority share in the production of capital goods (57 percent of sales), and accounting for a significant share of non-durable consumer goods and intermediate goods (43 percent and 37 percent respectively). They also dominated manufactured exports. The private national firms were complementary to the TNCs, dominating civil construction, the financial system, services, agriculture, and mining.

The modernization of the economy was achieved, but the authoritarian regime was not a necessary condition for its execution. The three fundamental structural changes which occurred simply accelerated a process which had already been initiated in the 1950s. The first of these changes was the dislocation of the dynamic axis of the economy from agriculture to industry. Brazil definitively ceased to be an essentially agricultural country, as the industrial share of GDP grew from 25 percent to 38 percent between 1960 and 1980 (Table 4.3).

The second structural change was the definitive displacement of the export sector as the impetus to growth: between 1947 and 1979 the ratio of exports to GDP declined from 14.8 to 6.7 percent. This meant that Brazil became a "closed economy," in the sense that the productive base came to depend upon the internal market (Serra 1982). Simultaneously, the state helped to expand the agro-industrial complex in the center-south by providing ample credit for producing

Table 4.3. *Evolution of economic sectors, 1960–1980 (% of GDP and economically active population (EAP))*

	1960		1970		1980	
	GDP	EAP	GDP	EAP	GDP	EAP
Agriculture	22.6	53.7	17.1	44.3	7.6	29.9
Industry	25.2	13.1	29.5	17.9	38.1	24.4
Services	52.2	33.2	53.2	37.8	54.3	45.7

Source: IBGE 1989.

agricultural raw materials for industry and export, for releasing workers from agriculture, and for expanding the agricultural frontier in the interior. Finally, the social structure was altered. A "demographic transition" was effected. The economically active population in the secondary and tertiary sectors grew, the middle class diversified and a substratum of a mobile working population was formed which could attend to the needs of the new investment poles in the cities and the frontier.

Social inequality worsened simultaneously with the continuation of large margins of absolute poverty. In 1974–75, one-third of Brazilian families, representing some 30 million inhabitants, lived below the poverty line. The modernization of agriculture was extremely uneven, with negative implications for food prices and the real income of rural workers. Although the capital-goods sector may be more ample and integrated in Brazil than in the other Latin American countries, it is relatively backward, revealing the insufficiency of national technology.

The economy continued to have major problems in its extreme dependence on petroleum imports, its chronic and increasing inflation rates, and the rising internal debt which the official banks contracted to finance large projects. Politically, these problems provoked cracks in the pact of power, raising the potential for conflict between different sectors of the bourgeoisie, particularly between the dominant multinational financial capital and private national industrial capital, and between these sectors and the state enterprises, as well as with the federal government and between the different governmental spheres of power.

In 1979, the second oil shock and the elevation of interest rates in the international market dealt a hard blow to the economy, and economic policy responded by containing spending, eliminating incentives, and reducing financing in all sectors. The external debt increased and became uncontrolled.

In the 1980s, then, the Brazilian economy moved forward within the context of crisis and reinvigoration of the commercial balance through increasing exports, and under the questioning of the regime's political legitimacy – a process in which the degree of urbanization resulting from the economy's own expansion and the state's intervention played an important role.

Territorial strategic planning and national integration

The concept of making the territorial structure adequate for the proposed industrialization was already present in the *Plano de Metas*; although it was in the First National Development Plan (1972–74), and especially in the Second National Development Plan (1975–79), that the project for organizing the national territory according to the logic of the geopolitical project was consolidated.

The policies for integrating the national territory were rapid and combined action aimed simultaneously at completing its occupation, incorporating the center west and the Amazonian "island"; expanding and modernizing the national economy, linking it to the international economy; and extending the state's control over all activities and all locations. These policies also had the function of legitimating the state. Once again, emphasis was placed on the idea of strengthening the fatherland, the nation-state being "advertised" in the media. Once again territorial integration was used as a symbolic resource for building a "Great Brazil." And in this context, the occupation of the Amazon took on the highest priority. In other words, the policies for integrating the territory sought to remove material and ideological obstacles to modern capitalist expansion.

The scientific–technological developments gave the state technical and conceptual capacity to deal with space on a large scale. A new spatial technology of state power developed, imposing on the national territory a powerful *double-control network which was technical and political* (corresponding to the government programs and projects) which we call "the planned network." It was manifest principally in: (a) the extension of all kinds of networks – transport, urban services, communication, information, institutional, banking, etc.; and (b) the creation of new territories superimposed on the existing political–administrative divisions, managed by state institutions, to whom investments were channeled (Becker 1988, 1989).

Incorporating the already existing tendencies of the economic and social reality, the government net involved the territory as a whole and acted on various scales, seeking to mold the territory according to the model of a space of discontinuous and connected valorization through "concentrated deconcentration" of industries and services. The elements of the "planned network" are most apparent in the policies of urbanization, regional development, growth poles, and occupation of the Amazon.

Urbanization as a strategy and the formation of national networks

In the past, the needs of the city had been a motive for the beginning of industrialization by import-substitution while the growth of the cities provided economies of agglomeration. Now, industrial growth had to be supported by uninterrupted urbanization. The urban nuclei assumed a new significance as the logistic basis for accelerated modernization: they assured the intensification of capital and labor mobility, and were the loci for tertiary activities sponsored by the government, state institutions, and the diffusion of controlled information.

The territory's urbanization became, then, a strategy for the country's development. After 1973, a spatial strategy related to territorial organization and deconcentration appeared explicitly in the official planning records with the creation of the National Commission for Urban Development (CNDU). In the Second National Development Plan (1975–79) there was a break with the option of local planning and the urban theme began to be treated from two angles: the internal organization of the urban centers, and the urban network. Policies of urbanization involved a group of strategies for developing capitalism in the country (Figure 4.5).

To these should be added the policies of expanding the built environment (Schmidt 1983), which were not restricted to the limits of the city, since they extended to the entire national territory. They sought to increase the speed of the system, which involves the growing importance of internal transport and communications and the unification of the national market. To this end, there were sectorial plans for heavy investments in fixed capital, represented by hydro-electric projects, airports, dams, plants, ports, etc., and the expansion of ground, marine, and air transport, the means of communication, the energy network, oil pipelines, irrigation canals, which criss-cross the country today in all directions. In sum, it was an expansion of the networks of territorial linkage, that is, of urbanization in its most ample sense.

Regional and growth pole strategies

The institutionalization of the macro-regions was the first step in this strategy of regionalization by the Ministry of the Interior (MINTER). By the second half of the 1960s, following SUDENE's experience, MINTER had created the regional superintendencies – for the Amazon (SUDAM), the center west (SUDECO) and the south (SUDESUL). This strategy sought to neutralize the regional oligarchies through new pacts and to organize the bases for modernization. The regional elites were co-opted at the same time that the federal system of fiscal incentives promoted the transfer of capital to the periphery. These incentives consisted of exemptions from federal and state taxes and suspensions of import tariffs for machinery and equipment required for executing the new projects. The new undertakings, whether national or foreign, also enjoyed special financial incentives through subsidized credit.

After the crisis of 1973, the governmental strategy became more selective, no longer acting on the macro-regional scale and beginning to focus on the sub-regional level through the strategy of implanting growth poles. Few countries of the world have taken the ideas of Perroux as far as Brazil did with the creation of these development poles. The ideology of development poles showed itself to be the most adequate model for territorial organization proposed by the authoritarian state, since it involved the creation of privileged locations, from the perspective of capitalist accumulation, capable of interlinking the national and international circuits of financial and mercantile flows (Egler 1988).

The Second National Development Plan (II PND) emphasized the comparative advantages of Brazil's various regions and encouraged the development of regional specializations. Growth-pole programs were placed under the administration of the macro-region superintendencies, with the exception of the southeast (Figure 4.6).

The occupation of the Amazon

The integration of the Amazon was promoted to the highest priority for reasons of accumulation and legitimation. Its occupation was thought capable of promoting internal and external "geopolitical equilibrium," offering a complete solution for the problems of social

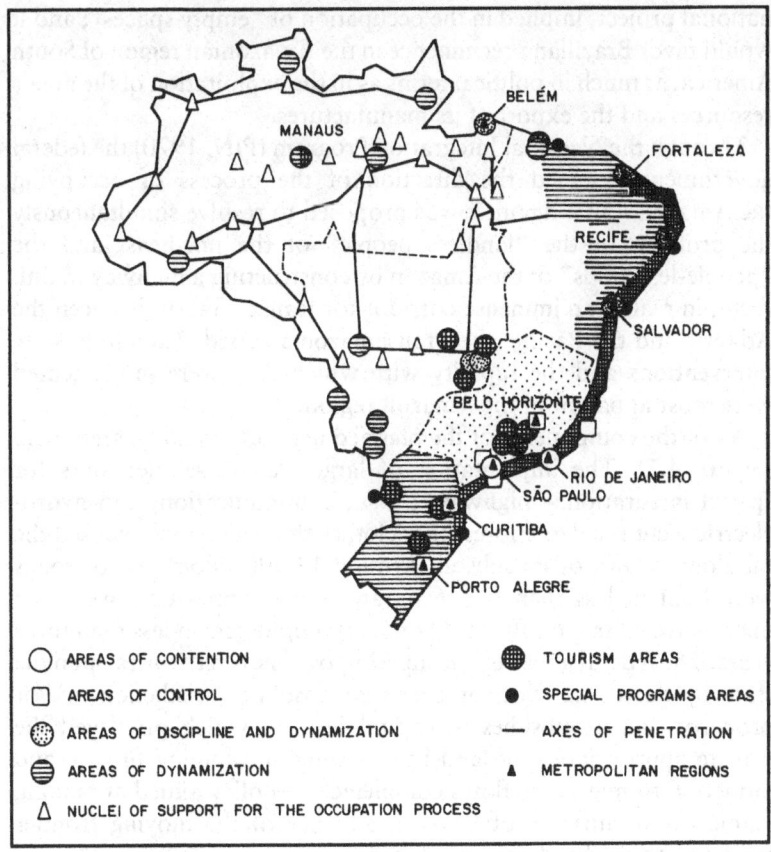

Figure 4.5 Urban development policy, 1975–1979.
Source: II National Development Plan 1975–1979.

tension in the periphery and for growth in the center, as well as serving to increase Brazil's prominence in South America (Becker 1982). Actually, territorial occupation, this time on a gigantic scale and at a newly accelerated rhythm, was expected to sustain the path of authoritarian modernization. It would help to avoid agrarian reform – necessary in the face of the modernization of agriculture and the consequent release of labor power – by pushing small producers to the interior, and encouraging migration from areas with social tensions, particularly from the northeast and the large metropolises; it would also assure latifundio reproduction. Occupation of the Amazon would also give substance to the ideology of "Nationalizing the Territory" as a symbol of the construction of a national project, implied in the occupation of "empty spaces"; and it would favor Brazilian preeminence in the Amazonian region of South America, as much in political terms as in the exploitation of the area's resources and the export of its manufactures.

Through the National Integration Program (PIN, 1970) the federal government assumed the direction of the process of occupying the Amazon. Once again, it was proposed to resolve simultaneously the problem of the "landless people" of the northeast and the "people-less lands" of the Amazon by constructing a highway to link them, in reality an immense corridor for transportation between the Atlantic and the Pacific: the Transamazonian road. Therefore, state interventions and the rapidity with which they were implemented were most apparent in the Amazon region.

All of the components of the planned network could be seen there (Figure 4.7). The implantation of large transversal networks for spatial integration – highway, urban, communication, and hydro-electric – cut the dense forest that clothes the region and exposed the fabulous wealth of its subsoil. Around 12,000 kilometers of roads were built in less than five years and a communication system in microwaves ("tropodiffusion") of 5,110 kilometers in less than three years. The federal government superimposed new territories upon the state by decree in which it exercised absolute jurisdiction and/or property rights. Subsidies to capital flows – which privileged the private appropriation of land by ranching and mining firms – and induction to migratory flows completed the policy aimed at making viable the occupation of lands in advance of the moving frontier restricted to the borders of the forest.

At the beginning of the 1980s, the spatial strategy for the Amazon expressed the "Forced March" and the economic crisis. Regional policy executed by conventional bureaucratic agencies was replaced by the implantation of big projects for mineral exploitation which sought to sustain economic growth through exports. They were managed directly by state enterprises endowed with research and development centers and a techno-bureaucracy. In other words, this represented a new phase of trying to attract foreign investment and also to expand and transnationalize the state enterprises.

The strongest mark of this new strategy is the Grand Carajás Program (PGC). The role of the state expanded in order to be compatible with the new scale of resource mobilization that was projected. Government instituted a new sphere of power in 1980, the Interministerial Council of the PGC, jointly with the Planning Secretariat (SEPLAN), directly connected to the central government

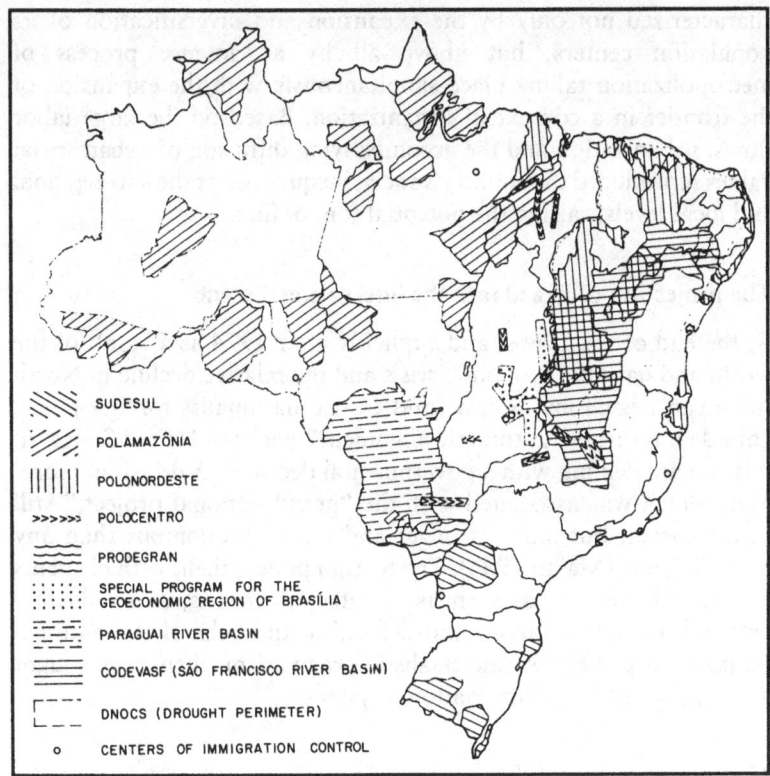

SUDESUL

POLAMAZÔNIA

POLONORDESTE

POLOCENTRO

PRODEGRAN

SPECIAL PROGRAM FOR THE
GEOECONOMIC REGION OF BRASÍLIA

PARAGUAI RIVER BASIN

CODEVASF (SÃO FRANCISCO RIVER BASIN)

DNOCS (DROUGHT PERIMETER)

o CENTERS OF IMMIGRATION CONTROL

Figure 4.6 Regional policy, 1975–1979.
Source: II National Development Plan 1975–79.

and creating a new territory of 90 million hectares, corresponding to 10 percent of the national territory. Second, it implanted the basic infrastructure for transnational production: the logistic global transport system, the hydroelectric network which generates the key input for producing alumina and aluminium, the railway network which favors exports, in addition to various urban nuclei. Third, the Companhia do Vale do Rio Doce (CVRD), the state holding that controls mineral exploitation in the country and is the world's largest exporter of iron ore, became the sole manager for the Carajás Iron Project including mines, a railway (900 km), a harbor, controlling a territory of 2 million hectares inside the Grand Carajás Program.

In consequence of the social dynamic and state intervention, labor and capital movements were intense, transforming Brazil's regional center–periphery structure. Brazil became an urban country, characterized not only by the expansion and diversification of its population centers, but above all by an intense process of metropolization taking place simultaneously with the expansion of the frontier in a context of urbanization. Based on the large labor flows, urbanization and the accompanying diffusion of urban social values reproduced the country's social inequalities at the sub-regional and local levels, raising the potential for conflict.

The projection of Brazil into the international scene

At the end of the 1960s, and explicitly after 1974 as a result of the world and national economic crisis and the relative decline in North American hegemony, a new kind of neo-nationalist foreign policy unfolded: refusing "automatic alignment" with the United States. In this, we are dealing with a governmental decision. A decision, more-over, which was associated with the "grand national project," still authoritarian, but much stronger and more autonomous than any preceding one (Malan 1986). The foreign policy, then, with elements of internal policy, was seen as "strategic," that is, as irreducibly national in the long term, and defined in the light of the national geopolitical project. A universalistic pragmatism, then, was sought with operation in the international system.

The foreign policy of the authoritarian regime

Between 1964 and 1974 the geopolitical theses of the ESG informed the strategy of the Ministry of Foreign Relations (also known as Itamaraty). Brazil grounded its foreign policy on a strong bilateral alliance with the United States, based on the ideas of ideological blocks and "loyal bargaining power." It sought to assume a hegemonic regional role in exchange for its alliance with the United States. Two efforts marked this period, aimed at assuming a larger role, first, in Latin America and, second, in the South Atlantic.

The first of these efforts was a Brazilian diplomatic offensive in Latin America, through negotiating bilateral relations which reflected US and Brazilian interests. In the ideological area, the foundations of Itamaraty's diplomacy were a "doctrine of fence-building" and a "thesis of preventive ideological warfare." The latter was apparent in the secret logistic support for military coups, as in Bolivia, and in the repression of leftist movements, as in the quasi-invasion of Uruguay. In the economic area, the Brazilian strategy sought to increase its own political influence and neutralize that of

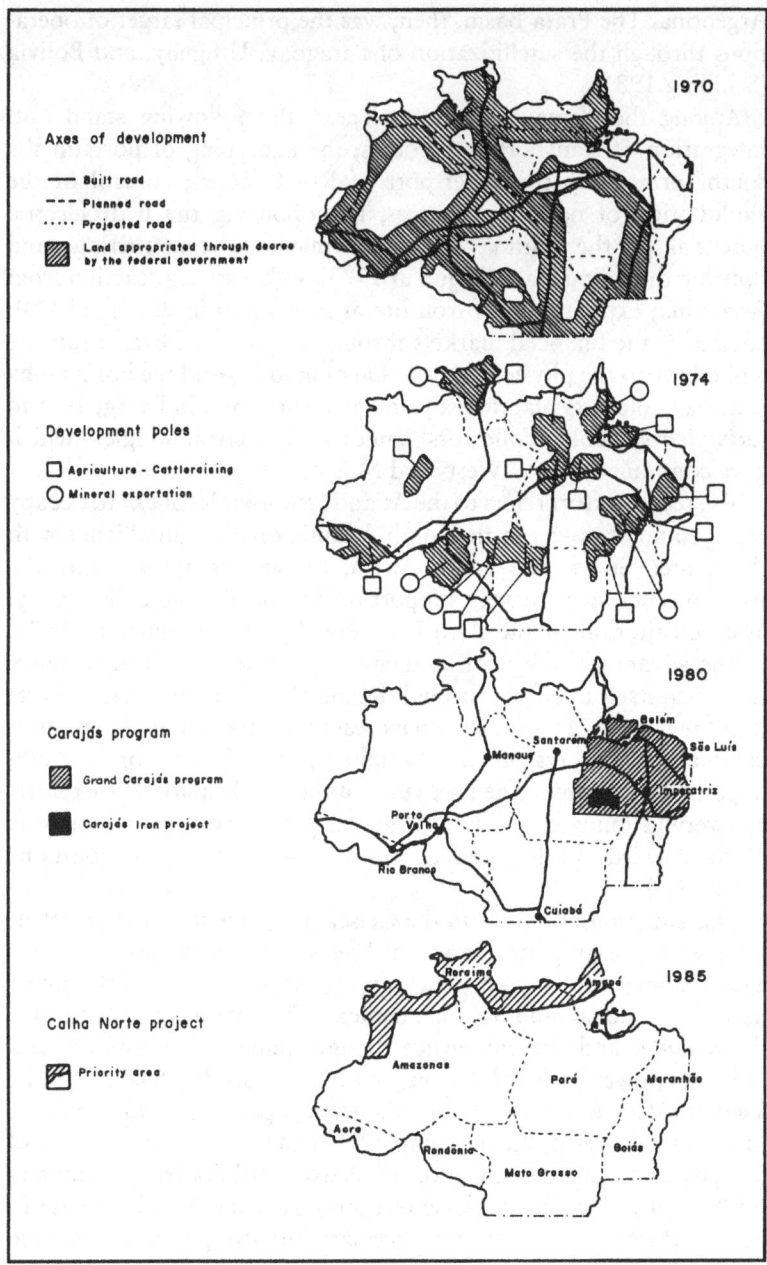

Figure 4.7 Legal Amazon: occupation policy.
Source: Becker 1990, adapted from various publications.

Argentina. The Prata Basin, then, was the principal target of operations through the satellitization of Paraguay, Uruguay, and Bolivia
(Schilling 1981).

Among the elements of this process, the following stand out:
integration of transportation routes; the equipping of ports in the
south (principally the super-port of Rio Grande); control of the
exploitation of natural resources, monopolizing the hydroelectric
potential of the Paraná River by financing, constructing, and
utilizing the Itaipu Dam (begun in 1973) with a strong reaction from
Argentina; exploitation of iron ore at El Mutum in Bolivia (1974);
control of the financial markets through the Bank of Brazil; turning
a blind eye to the physical occupation of land beyond the border – by
Brazilian colonists planting soy and acquiring land in Paraguay, and
seringueiros (rubber collectors) entering the Bolivian jungle – which
gave continuity to the "Westward March."

The second effort refers to the Atlantic frontier. It sought to occupy
the vacuum of power in the South Atlantic through an alliance with
the United States and South Africa, as well as approaching the
African economies through support for Portugal's colonialist policy.
The constitution of the Afro-Luso-Brazilian Community in 1971,
taking advantage of historic and cultural ties, permitted the country
to launch itself in Africa through financial and commercial projects
(Schilling 1981). The attempt to increase the export of manufactures,
conquering markets in the periphery, is evidence of a "sub-
imperialist" strategy. The first years of the 1970s marked, therefore,
an overwhelming tendency for Brazil to pull away from the interests
of the "Third World" and to abstain on international questions
(Table 4.4).

This situation altered with the Geisel government which, in 1974,
initiated a foreign policy known as "responsible pragmatism." Once
again, changes in foreign policy expressed alterations in the international, regional, and national context. The world economic crisis,
the *détente*, and the emergence of new centers of economic and
political power reduced US hegemony, decisively shaken by the
Vietnam War. A power vacuum, therefore, opened in Latin America.
In the national plan, Brazil strongly resented the rise in oil prices and
the power of the Organization of Petroleum-Exporting Countries
(OPEC). It had reached a level of economic growth and increase in
power which made it not only necessary but also possible to pursue
a more vigorous foreign policy, with greater capacity for negotiation
and resistance to the pressures of the international system.

Three elements stand out from the policy of "responsible

Table 4.4. *Destination of Brazilian exports, 1969–1974*

Country or area	In millions of current $US		% increase	
	1969	1974	Exports	Accounted for by area
Latin America[a]	254	918	261	11.8
Bolivia	4	82	1,950	1.4
Paraguay	7	98	1,300	1.6
Africa	24	417	1,638	7.0
Angola	0.3	6	2,900	0.1
Mozambique	0.2	6	1,900	0.1
EEC countries[b]	782	2,434	211	29.0
United States	760	1,737	128	17.3
Total[c]	2,311	7,950	244	

(a) Latin American Free Trade Association; (b) includes exports to Britain; (c) because individual totals are included only for selected areas, individual figures do not add up to the quoted total figure.
Source: Evans 1979.

pragmatism." First, Brazil pulled away from the United States militarily and commercially. Already, in 1969, Brazil refused to sign the Nuclear Non-Proliferation Treaty. It was in 1974, however, that Brazil adopted a position against "automatic alignment" with the United States, based on the concept that economic interests overruled ideological ones, requiring their defense wherever they might be. The second element was a gradual interaction south-to-south. The third element was the disconnection of diplomacy from ESG's geopolitics; the diplomatic discourse rejected Brazil's "status" as a hegemonic power in order to strengthen the country's independence in the international context. Brazil sought, therefore, to act on two different levels. On the one hand, it tried to expand its horizon of options and its role in the world context. On the other hand, it also sought to affirm its presence in South America, a region considered to be its natural area of interest.

As for its insertion into the global system, Brazil's activities involved relations with almost every part of the globe. First, it recognized the governments of Angola, Mozambique, and Guinnea-Bissau, thus disconnecting itself from the Portuguese policy and occupying its own space in the former colonies. Angola came to be the principal object of Brazilian designs, through PETROBRAS' participation in exploration for petroleum and its subsidiary INTERBRAS' sale of capital goods. Second, Brazil reopened relations with the People's Republic of China in 1974, even against the resistance of extremist sectors of the military. Third, Brazil approached the Arab countries, in detriment to its relations with Israel. In this, it sought to attend to its vital needs for petroleum, to attract Arab capital, and to open new markets – such as offering civilian services and selling armaments. Fourth, Brazil signed a nuclear treaty with the Federal German Republic in 1975 and, in 1977, broke its military agreement with the United States which had been in force since 1952. Finally, it expanded commercial ties with Eastern Europe and Japan and constantly countered North American protectionism in international organizations (GATT and UNCTAD).

A second level of operation was regional, where Brazil sought to affirm its position on the continent through mutual cooperation, trying to undo the image of imperialist expansion created in the earlier period and presenting itself as merely one member of the Latin American community. Two changes stood out in this policy: (a) a change in the style of conducting Brazilian diplomacy, giving priority to bilateral negotiations, even without abandoning active participation in regional entities; and (b) a change in the goals of continental relations, privileging not merely the Prata Basin but now also the Amazon Basin. This change sought to obtain alternative sources of petroleum from Venezuela, Ecuador, and Peru, as well as coal from Colombia. It aimed at increasing the sale of Brazilian manufactures – vehicles, domestic appliances, agro-industrial products, as was the case with Venezuela, and to reduce the impact of the Treaty of Cartagena (1969), known as the Andean Pact, which obstructed the access of Brazilian products to these markets. Finally, it sought to take Guyana and Surinam out of their isolation.

The Treaty of Amazonian Cooperation (TCA), proposed in 1977, aimed at increasing regional development and ecological protection, under the premise of refuting any kind of attempt at external meddling in the region. This completed, then, the framework of diplomatic action on the continent. It was no longer restricted to the Prata Basin, although scarce resources and external and internal divisions made it difficult to institutionalize the treaty and effect the integration of the Amazon Basin.

The limits of a regional power

The most conspicuous characteristic of the recent Latin American experience is diversity. While Argentina and Chile deindustrialized and Mexico desubstituted imports, Brazil, already the major

Table 4.5. *Gross domestic product per capita, 1950–1980*

Country	In 1975 prices (dollars)					
	1950	1960	1973	1980	1985	1950/1985
Argentina	1,877	2,124	3,045	3,209	2,719	1.45
Brazil	637	912	1,624	2,152	2,072	3.25
Chile	1,416	1,664	2,108	2,372	2,135	1.51
Colombia	949	1,070	1,536	1,882	1,878	1.99
Ecuador	638	758	1,190	1,556	1,448	2.27
Mexico	1,055	1,401	2,170	2,547	2,436	2.31
Peru	953	1,200	1,740	1,746	1,563	1.64
Uruguay	2,184	2,501	2,653	3,269	2,727	1.25
Venezuela	2,127	2,839	3,468	3,310	2,671	1.26

Source: Balassa, Bueno, Kuczynski, and Simonsen 1986.

industrial power in the continent, consolidated its leadership (Hirschman 1986). Brazil detached itself from Latin America. It accounts for one-third of the population and of the gross domestic product of the entire region, showing the best performance in the GDP per capita (Table 4.5).

The amplification of Brazil's power owed itself to the national project for modernity formulated and executed by the state. Internally, a key part of this project was the intentional policy of increasing national capacities and decreasing technological dependence on the United States through accords with Japan and Germany, and the realization of the country's own path for the computer industry. On this point, the role exercised by the military in the experiences which were most successful for top-level technology should be emphasized, following its historical tradition in the struggle for steel and petroleum. With all their limitations, these initiatives were essential because the national industrial bourgeoisie, due to its dependence on foreign interests and its own ideological weaknesses, was incapable of defining its own project for industrial policy (Coutinho 1987).

In the sphere of foreign policy, Brazil exercised its capacity for negotiation, permitted by the fluidity of international events, initially in relation to Latin America and later at the level of foreign relations with Africa, extending its economic, financial, technological, and political influence.

Brazil's contradictions as semiperiphery are clearest in international trade. The direction of commercial flows still places it among the peripheral nations, which trade more with the developed

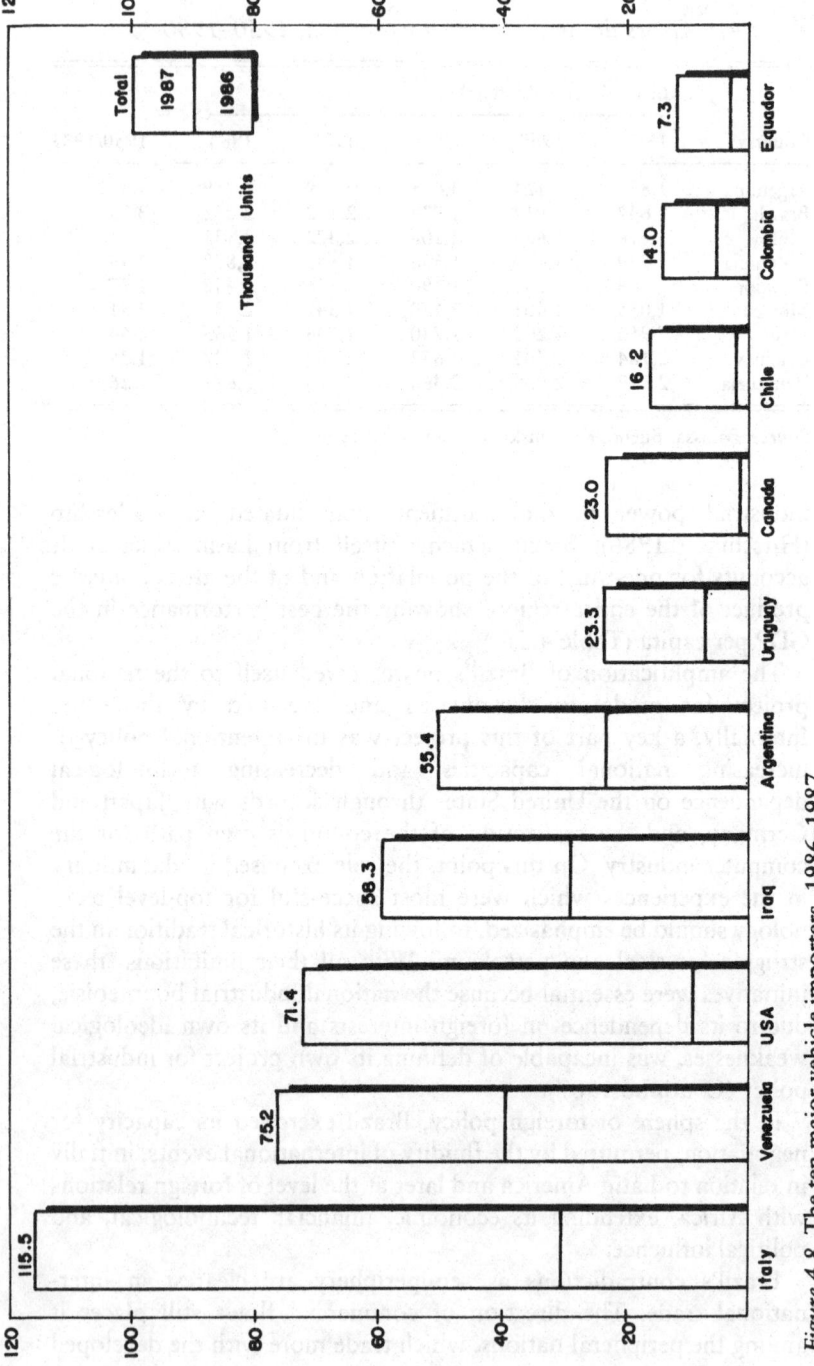

Figure 4.8 The ten major vehicle importers, 1986–1987.
Source: Brazilian Association of Automotor Vehicles (ANFAVEA).

Table 4.6a. *Structural change of Brazilian exports, 1965–1982*

Year	Primary (%)	Semi-manufactured (%)	Manufactured (%)
1965	82.1	9.7	8.2
1970	75.5	9.2	15.3
1975	59.4	10.0	30.6
1980	42.7	11.8	45.5
1982	41.1	7.4	51.5

Source: Bank of Brazil.

Table 4.6b. *Structure of Brazilian exports, 1981–1989 (millions of US$)*

	1983	1985	1987	1989
Total	21,592	23,359	26,056	33,999
Primary	8,535	8,538	8,022	9,599
Coffee	2,096	2,369	1,959	1,610
Iron ore	1,465	1,658	1,615	2,233
Soy bran	1,793	1,175	1,450	2,136
Others	3,181	3,336	2,998	3,620
Manufactured	13,057	16,821	18,034	24,400
Transportation equipment	991	1,694	2,780	3,886
Machinery	828	1,590	1,634	1,833
Iron and steel	1,151	1,357	1,071	4,058
Others	10,087	12,180	12,549	14,623

Source: Fundação Getúlio Vargas, Conjunture Econômica 44(11), 1990.

nations than with their neighbors. Brazil's major clients and suppliers are still the United States and Europe (with the exception of the supply of petroleum from the Middle East). Nevertheless, three elements reveal Brazil's status as semiperiphery. The first is the nature of Brazilian trade. After the recession at the beginning of the 1980s, Brazil stood out from its Latin American neighbors with an extraordinary rebound in exports. Manufactured goods accounted for a growing share of these exports (Figure 4.8), increasing from 30 percent of total exports in 1970 to 45 percent in 1980 (Table 4.6).

The second element is the opening of markets and the diversification of trade partners. Between 1970 and 1980, the share of Brazilian exports purchased by the industrialized nations fell from 72 percent to 45 percent, while the peripheral nations increased their share from 28 percent to 45 percent. The expansion of manufactured exports depended upon the expansion of trade with Latin America,

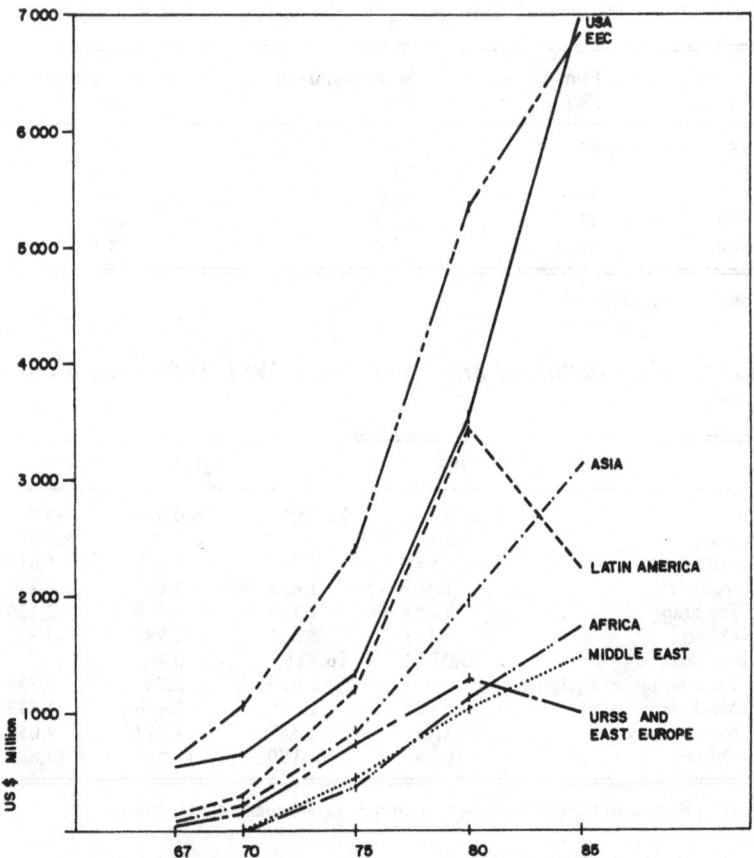

Figure 4.9 Increase in Brazilian exports by destination, 1967–1985.
Source: Bank of Brazil.

Africa, and the Middle East, expressed in the foreign policy which Brazil adopted, although these exports also increased to the United States and Japan. The universalistic tendency was accentuated in the 1980s with a new maneuver: that of transactions among the semi-peripheral countries. East Asia – the "tigers" and China – became the fourth major trading partner for Brazil (Figure 4.9). In 1988 they absorbed 10 percent of Brazil's exports of iron ore, steel products, chemicals, petrochemicals, and textiles.

The third element to emphasize in trade is the opening/closing of the Brazilian economy. Between 1947 and 1980, the share of GNP represented by exports fell from 14.8 percent to 7.6 percent; imports

Table 4.7. *Percentage of total investments and reinvestments in*
Brazil by country of origin, 1970–1982

Countries	1970	1974	1978	1982
United States	42.2	33.6	27.8	31.2
Germany	10.8	11.8	15.3	13.8
Britain	8.8	6.6	5.4	5.1
Japan	4.5	9.9	10.2	9.2

Source: Central Bank of Brazil.

also declined from 13.7 percent to 7.0 percent of GNP over this period. Foreign trade no longer constituted the most significant element of the economy, representing in 1985 only 15.6 percent of GNP. This signified a notable level of economic "closing," thanks to the expansion of the country's internal market. This closing has a cyclical trajectory, having occurred most intensely between 1950 and 1965, and again after 1974 (Serra 1982). This brings up the ambivalence of the Brazilian pattern. The state guaranteed a captive market for firms, such that it became a "Cartorial State," which keeps a weighty segment of the market artificially protected by subsidies.

With respect to the degree of control of its economy by foreign investment, Brazil's position is not so clear. Brazil was the peripheral nation which received the most direct investment from foreign companies. Among the 500 largest firms in Brazil, 50 percent are foreign. Today, multinational capital accounts for almost 30 percent of Brazil's industrial production, 27.4 percent of its exports and 17.9 percent of its imports. But this type of "dependency" must be put in context. The composition of investments diversified, reducing the shares of the United States and Britain in favor of those of Germany and Japan (Table 4.7).

Furthermore, foreign capital is today an integral part of the Brazilian economy; it ceased to be an external force operating locally, as occurs throughout the world, including the United States and the EEC. As a counterpart to this, the Brazilian state enterprises transnationalized, signifying Brazil's participation in other economies and the affirmation of the nation-state (Becker 1988). PETROBRAS (petroleum), which diversified, by 1973 was among the 200 largest non-American industrial firms in the world and in 1983 it kept this position, figuring as a multinational and becoming an instrument of Brazilian foreign policy. EMBRAER (aeronautics) exports

50 percent of its production today to the developed countries, where it has two affiliates. CVRD (iron ores) has diversified and maintains interests in the United States and in Asia. On a planetary scale the networks of the Bank of Brazil (finance) and VARIG (airline) have also extended.

The multinationalization of the state enterprises is associated with another element to be considered in evaluating Brazil as a regional power: its technological and financial capacities. This is also displayed in the operation of the federal government and the private national firms of large engineering projects, whether or not they are associated with governmental projects.

The major projects of the federal government in this area have occurred above all in Bolivia and Paraguay, as a result of their proximity and of Brazil's continuing rivalry with Argentina. The Brazilian government financed and guaranteed the purchase of part of the production of the first steel plant in Bolivia in exchange for Bolivian natural gas and petroleum. Through the extension of the road and rail networks, Brazil also co-opted the rich Bolivian region of Santa Cruz. As for Paraguay, in addition to the construction of the Itaipu Dam by a binational organization – which in reality was financed, constructed, and is utilized by Brazil – the "Friendship Bridge" was built, linking the roadways between Assunción and the Port of Paranaguá in southern Brazil. The addition of fiscal facilities in this port co-opted the commerce of the Paraguay River. In consequence, the mobile Brazilian frontier surpassed the country's territorial boundaries, creating a population of 7,000 *brasilguaios* on Paraguayan soil.

Large-scale engineering operates through the provision of services, accompanied by technical formation, labor training, and the sale of equipment. It is realized by contractors who compete with the developed countries in the exploitation of natural resources and the construction of dams and highways through bilateral accords. Between 1966 and 1980, 68 percent of the contracts for industrial projects which were exported by Brazil went to Latin America (BID 1982), but this activity extended also to Africa and the Middle East.

Brazil still affirms itself as an original power by its operation on the political plane through a diplomacy joined to a strategy of "grand regional zones" in South America, with that of "cultural affinities" in Portuguese Africa. It acquired, in this way, a certain "overt power" (Taylor 1985) which allows decisions to be steered along certain directions in the continent.

Meanwhile, the authoritarian regime, ruled by decisions from the top, ignored any criticism concerning the growing vulnerability implied in the process of external indebtedness. The external debt of Brazil was one of the bases for a policy of accelerated modernization, but was also an integral part of the wider phenomenon of structural changes in the world economy, that is, of the globalization of the economy and the vigorous loan-pushing developed by the large banks during the 1970s. While the central economies suffered the oil shocks, the banks reaped large profits from recycling petro-dollars. They had problems resolving where and how to invest while their own economies were in recession; therefore they directed loans to peripheral countries. As for Brazil, this loan-pushing coincided with government decisions to maintain the economy in a "Forced March." As a result of this combined movement, between 1973 and 1978 the debt quintupled, increasing from US$6,155 million to US$31,616 million, although it corresponded only to 32 percent of Brazil's GDP until 1981.

The borrowing corresponded to the construction of the most recent large blocks of infrastructural investment and to building the industrial capacity necessary for the geopolitical project. But the country was forced into recession when the accelerated rhythm of these investments met conditions which were steadily more vulnerable to the internal and external financing situation, the international banking embargo, and pressures from the governments of the advanced nations. With the continuing rise in international interest rates the debt emerged as a serious structural problem, a constraint to be confronted in future decades (Coutinho 1987).

Strategic constraints are also important. According to Cavagnari Filho, a colonel who opposes the geopolitical discourse of "Brazil as a Power," one needs to reconsider the absolute indicators of National Power in the light of qualitative indicators. Brazilian growth in recent years registered considerable progress in relation to the under-developed countries, but it did not manage to reduce significantly the gap which remains in relation to the industrialized democracies. Brazilian GNP is about 8 percent of the US GNP and 50 percent of the GNP of Britain. Brazilian investment in research and development (0.7 percent of GNP) is about 2 percent of North American investment and 15 percent of British, and Brazil's military force is much less than that of the advanced countries.

Considering the military domination of the United States, and both the absolute quantitative and qualitative indicators, Brazil would be a *medium power*, whose strategic sphere is limited to South America. The position which it occupies in the hierarchy of world power as the first country of South America concedes it, by extension, the status of greatest regional power. Its actual strategic capacity has sufficient reach to operate within this area in defense of its vital interests, but does not confer upon it the degree of autonomy that would be desirable to develop strategic initiatives (Cavagnari Filho 1987).

In summary, considering the multiple dimensions of power, and the limits imposed by the increasing foreign debt, it is possible to define Brazil as a regional power, restricted in strategic terms to South America, with political influence in Latin America and Lusophone Africa, and with economic operations which extend to the Middle East.

The national question, redefined, has passed today on the internal plane to the social question and to full nation-building; and autonomy, on the foreign plane, has shifted to the technological question and debt. Along with inflation and internal debt, these comprise the challenges which Brazil will confront in the next decades, and whose solution will also depend upon the trajectory of the United States and the path of the world-economy.

5

The legacy of conservative modernization and territorial restructuring

Brazil became modern by an authoritarian path and the geopolitical project, elaborated and directed by the Armed Forces, has left a profound imprint on society and space. Brazil's GNP grew until it reached its current position as the eighth largest in the world, its industrial base attained a high degree of complexity and diversification, its agriculture displays indications of immense technification and dynamism, and an extensive network of services interconnects almost its entire national territory.

Nonetheless, the majority of the Brazilian population has not benefited directly from this economic growth. Brazil inaugurated modernized poverty. This is not primitive poverty, but rather a poverty which is illuminated by the small windows of television screens dispersed throughout thousands of houses, shacks, and shanties. Connecting the rich, the well-off, and the poor within an illusory and utopian world of serials and programmed news, the electronic ideology of television performs a role, unique in the world, as a political and social instrument for forming opinion during the dictatorship, and even afterwards.

Modernity exists with this poverty in a complex web. How does one explain the skills of thousands of mechanics spread along the vast highway network, who are capable of servicing an extremely diverse fleet of vehicles even though few have formal training or even learned to read and write? This is not a case of "traditional" and "modern" separated by a clear dividing line. It is misleading to describe Brazil as "two Brazils," or as a "Belindia" – a Belgium beside an India. More than this, Brazil is a hybrid structure, ambivalent, unstable, yet very dynamic. This is the legacy of

117

conservative modernization, which will be analyzed in this chapter at the social, economic, and spatial levels.

Modernized poverty

Conservative modernization generated a specific kind of modernized poverty. The social problem of the semiperiphery is revealed in an enormous disjunction between the expanding network of collective services and equipment and the precarious social conditions of the nation. The authoritarian regime tried to extend social policy massively, thereby degrading the quality of services. "The managerial, operational, and administrative problem of social policy is the forgotten space of the state apparatus" (Lessa 1990).

With all this, the planned network generated unpredictable effects – externalities of the model – signifying deep structural changes, such as the "demographic revolution" and social fragmentation. Furthermore, the social dynamic evades state regulation. The officially regulated structure is counterposed to a subterranean society which is "parallel" or "non-official," creating its own rules and specific forms of resistance.

The demographic explosion that never was

The declining birth rate is perhaps the most important transformation facing Brazil today, with implications which are as yet unknown. The lack of public knowledge about this new fact derives from its unprecedented character relative to previous trends, from the exclusion of information to a restricted circle of specialists, and from the international campaign against the threat of a "demographic explosion" (Martine 1989).

The demographic transition – moving from high to low levels of fertility and mortality – is distinguished in Brazil from the classical transition of the European countries by two basic characteristics. The first of these is the speed of Brazil's transition when compared to the traditional pattern. In a few decades, Brazil and other peripheral countries are completing a transformation which took one or two centuries to conclude in Europe. This speed is associated with a violent and surprising decline in birth rates, with consequences for the natural growth rate of the population. The highest level of natural growth in Brazil occurred during the 1950s and 1960s (2.9 percent per year) due to the decline in mortality associated with industrialization. From the end of the 1960s, however, birth rates

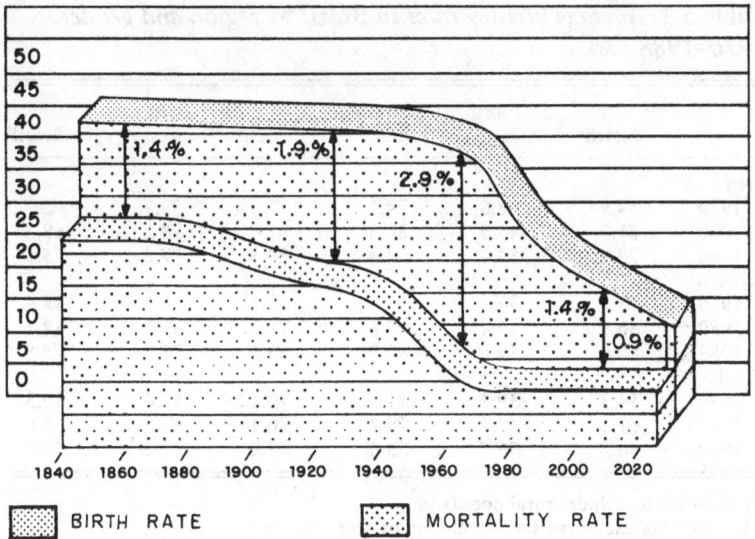

Figure 5.1 Demographic transition in Brazil (individuals per 1,000 inhabitants).
Source: Martine 1989.

began to decline (Figure 5.1). The census of 1980 revealed that fertility fell drastically throughout the country, as much in the rural areas as in the cities. This tendency has been confirmed in the 1980s. Between 1980 and 1984, the average number of children that a Brazilian woman would be likely to bear in her lifetime fell from 4.35 to 3.53, a decline of 19 percent. This decline was even sharper in the northeast. In consequence, the rate of demographic growth fell from 2.5 percent in the 1970s to approximately 1.8 percent in the 1980s, as currently estimated (Martine 1989).

The second characteristic which distinguishes the demographic transition in Brazil is that it is occurring independently of improvements in the material conditions of increasing sectors of the population. The declining birth rate was not the outcome of deliberate government policy, but accelerated modernization is nevertheless at its root. The declining birth rate is an indirect and unforeseen outcome of a series of government policies and expenditures for modernizing basic infrastructure and public services (Faria 1988; Hirschman 1986). In particular, policies for telecommunications, health, transportation, and education accelerated the diffusion of values, knowledge, new practices, and cultural attitudes which encouraged birth control at the same time that birth-control

Table 5.1. *Average literacy rates in Brazil by region and gender, 1970–1988 (%)*

	North[a]	North-east	South-east	South	Center west	Brazil[b]
Men						
1970	54.9	38.8	74.1	72.1	58.7	62.0
1980	61.2	45.9	80.8	81.4	68.2	69.3
1988	79.8	54.0	84.4	84.5	77.2	75.1
Women						
1970	53.7	39.6	69.0	68.1	55.1	58.6
1980	60.9	49.4	77.6	78.7	67.6	68.2
1988	80.6	59.4	82.9	83.2	77.8	75.8
Total						
1970	54.3	39.2	71.6	70.1	57.0	60.3
1980	61.1	47.7	79.2	80.1	67.9	68.8
1988	80.2	56.7	83.6	83.8	77.5	75.4

[a]Data of 1988 excludes rural population.
[b]Data of 1988 excludes rural population of north region.
Source: IBGE, Demographic Census of 1970 and 1980; IBGE, Domicile Sample National Research (PNAD) 1988.

methods became increasingly available. In a recent survey, 73 percent of married women between fifteen and forty-four years of age had used birth-control pills, 93 percent of whom purchased the product directly from a pharmacy without previously consulting a doctor (Martine 1989). In sum, the Brazilian population entered the era of the pill without leaving the era of misery.

The disjunction between economic and social indicators suggests that demographic and social behavior is not rigidly linked to economic cycles since it does not tend to reverse with temporary declines in income. This means that by the end of the century Brazil will have patterns of fertility and population growth comparable to those seen in the developed countries today.

The social state of the nation

More than half the Brazilians are poor. The nation's poverty is most apparent in the high rates of illiteracy, low incomes, and precarious living conditions. Almost one-third of the population over five years of age is illiterate, and this illiteracy is principally concentrated in the northeast. Even though school attendance has expanded, the primary schools have effectively failed. Brazilian children today attend on average only two-and-a-half hours of school a day

Table 5.2. *Income distribution in Brazil, 1970–1989*

%		1970	1980	1986	1989
Bottom	10–	1.2	1.1	1.0	0.6
	50–	14.9	12.6	12.5	10.4
Top	10+	46.7	50.9	48.8	53.2
	1+	14.7	16.6	15.2	17.3

Source: IBGE, Demographic Census of 1970 and 1980; IBGE, Domicile Sample National Research (PNAD) 1986, 1989.

and literacy rates are unequal by gender and by regions (Table 5.1).

Poverty is related to the low wages for unqualified labor which contrast with the high remuneration for technical and administrative services. Unequal income distribution has increased (Table 5.2).

Despite the reduction verified in regional inequalities – due to real gains registered by the north and the center west – income concentration grew. Almost 60 percent of the economically active population receiving some form of income (52.4 million individuals, of whom 35 million are men) earn no more than two minimum salaries, a level which defines the poverty line. Fully one-third of working Brazilians earn one minimum salary or less. The proportion earning below the poverty line is highest in the northeast and lowest in the southeast (Table 5.3a and 5.3b).

Furthermore, workers lack legal protection. Although employers are legally required to provide their workers with signed work cards, only a little more than one-half of Brazilian employees actually have these cards, which guarantee access to unemployment insurance, labor courts, and public benefits. This massive evasion of legal registration is one of the most shocking displays of violating the law. In this and related aspects, the worst situation presents itself in the northeast and among women. Precarious living conditions for families and infant mortality are corollaries of these situations. Inadequate and unequal access to public services reduces real income. For example, one of the worst problems affecting health is the absence of sewage facilities, a situation which is particularly serious in the northeast (Table 5.4). With the state's fiscal crisis in the 1980s, social services deteriorated in the extreme. Public education became steadily worse; violence intensified in the streets, neighborhoods, and homes; and the erratic and crowded system of collective transportation wore down the life conditions of workers who spent large parts of their days commuting.

Table 5.3a. *Gender income inequalities, 1981–1989*
(monthly median income in US$)

Gender	1981	1983	1985	1987	1989
Total	155.72	137.75	163.11	178.26	209.27
Men	250.48	219.83	258.59	277.61	327.64
Women	63.77	58.48	71.52	83.85	97.44

Note: Data excludes rural population of north region.
Source: Basic data from IBGE 1990b.

Table 5.3b. *Regional income inequalities, 1989 (monthly median income in US$)*

	Brazil[a]	North[b]	North-east	South-east	South	Center west
Total	209.27	217.02	107.10	265.28	212.44	235.34
Men	327.64	335.74	169.46	413.25	333.27	368.86
Women	97.44	108.86	48.97	126.83	95.12	106.39

[a]Data excludes rural population of north region.
[b]Data excludes rural population.
Source: Basic data from IBGE 1990b.

The Brazilian countryside is not comparable to rural areas in Africa, Asia, or even most of Latin America. The poverty related to Brazil's countryside is strongly linked to the urban centers. The largest share of Brazil's impoverished residents are in "rurban" areas, that is, urban centers with fewer than 20,000 inhabitants in which people depend as much upon temporary or seasonal jobs in agriculture as they do upon employment in the cities.

Information on wages, income, access to amenities, worker protection, and literacy shows that poverty is concentrated in the countryside, in the northeast, and among women. The data on social conditions, however, reveal little about the survival strategies created by the population to complement family income and resist absolute impoverishment. Some of these strategies are indicated by the declining rate of fertility and the increasing mobility of labor.

Labor-force mobility

Capital concentration and economic growth did not rest solely upon repressing wages; they were also founded upon the extraordinary

Table 5.4. *Water supply and sewage by regions (%), 1970–1986*

	Water supply			Sewage		
Region	1970	1980	1986ᵃ	1970	1980	1986ᵃ
Brazil	32.8	54.9	69.9	26.6	43.2	51.1
North	19.2	39.2	81.9	8.8	20.4	51.8
Northeast	12.4	31.6	47.4	8.0	18.2	28.2
Southeast	51.6	72.6	84.6	43.9	63.5	71.3
South	25.3	52.0	65.4	20.1	40.3	55.1
Center west	19.9	41.7	58.8	15.0	21.8	29.6

ᵃData excludes rural population of north region.
Source: IBGE, Demographic Census of 1970 and 1980; IBGE, Domicile Sample National Research (PNAD) 1986.

intensification of the historic mobility of the Brazilian workforce. The process of migration not only expanded poverty, but also created the new social fractions which make up the universe of capitalist society. At the same time, rates of turnover intensified while occupational multiplicity became more prevalent.

The modernization of firms and policies explicitly aimed at labor, as well as those with indirect effects upon it, induced this spatial and social mobility, which had two aspects. On the one hand, the dynamic spaces with new opportunities for employment and/or access to land attracted the labor force. This attraction was strongest in the southeast, the metropolises, and secondarily in the Frontier (center west and north) (Figure 5.2). On the other hand, agricultural modernization dislodged rural workers throughout the country, displacing the northeast from its historic position as Brazil's only region with substantial emigration. The government's subsidies for mechanization, best exemplified in soya cultivation, transformed the State of Paraná from a "moving coffee frontier" into a major exporter of labor in a single decade (1970–80). The increasing cost of land and differential access to credit conditioned a concentration of property which was effected through the violent expropriation of small producers whether squatters, sharecroppers, or small land-owners.

As a result, labor-force mobility increased on a national scale and fragmented the structure of social classes. This mobility is strongly associated with the formation of a new labor market with regional particularities. A diversified proletariat was formed whose primary category includes workers who move between rural and urban areas. In the areas where the labor market is most organized, such as São

Paulo, permanent rural wage-earners were transformed into temporary workers. These are *boias-frias* who live in towns and commute daily to work in the countryside. In less-capitalized areas, the traditional peasantry has adapted by becoming a hybrid proletariat/peasantry, selling their labor alternatively for rural and urban work depending on the season, and residing in urban areas. This process signified greater instability for the workers and greater exploitation of their work because it kept wages low, expanded the work day, and "freed" employers from their obligations.

A second type of proletariat corresponds to the employees of the formal and informal sectors of the large cities whose numbers respond to growth in these urban sectors. This includes the formation of a highly skilled workforce in the metal–mechanical industries of São Paulo. It also includes an enormous mass of employers and employees who make up a "parallel economy" which evades official regulation. This sector, which has been the object of very few studies, includes extremely diverse activities from peddlers to small factories.

The diversification and expansion of the middle class has been one of the most remarkable transformations of Brazilian society in the 1960s and 1970s. The middle class expanded in association with growth in the secondary and tertiary economic sectors and the state apparatus. The situation of the middle class is unstable in that it has consumption propensities which surpass the means available to satisfy them. The heaviest tax burden of the "official economy" falls directly upon this class.

The new meaning of urbanization

The major force behind the authoritarian modernization was an accelerated rate of urbanization, whose pace is among the highest in the world. This process acted as both an instrument and a product of governmental policy, and was also the result of unforeseen effects as well as spontaneous social adjustments. This is so because urbanization is the nexus by which Brazil is connected to the world-economy as a semiperipheral country. The urban nuclei are the *loci* of new institutions, circulation of goods, capital, and information. They are also the *loci* where the workforce expelled by the modernization of agriculture resides, circulates, and is resocialized to enter modernized poverty.

The structural transformation of the economy and society which characterizes the semiperiphery assumes concrete form and clearest

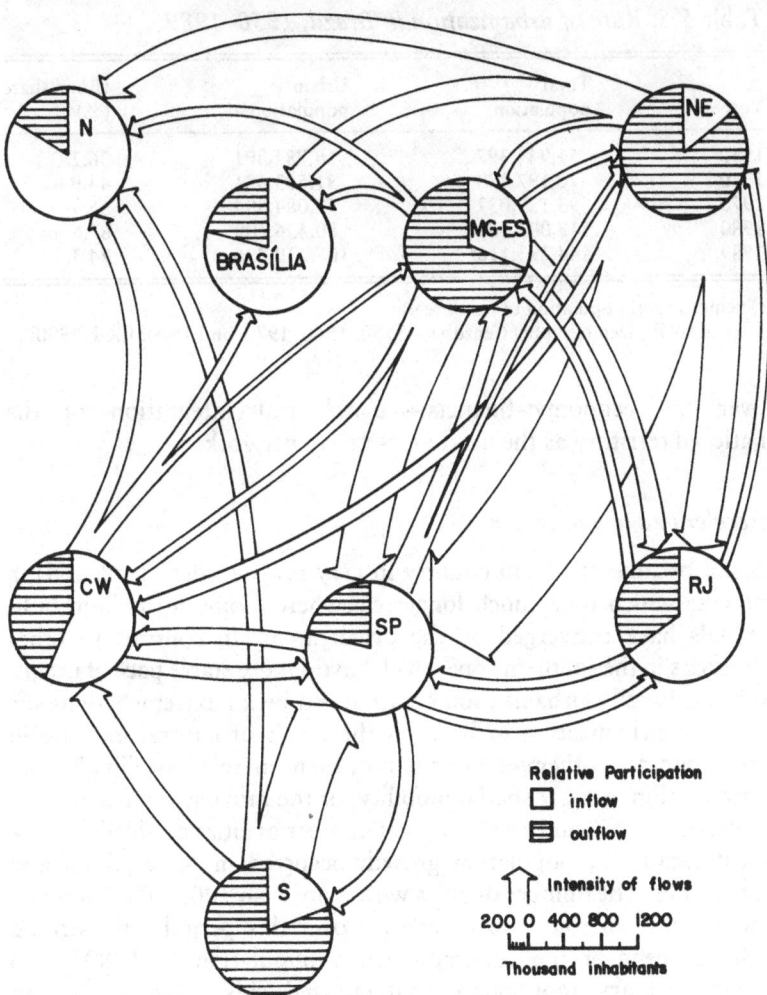

Figure 5.2 Migratory flows in Brazil, 1970–1980.
Source: Ablas and Fava 1984.

expression in the world city, São Paulo, which has become one of the world centers for capital accumulation and control. São Paulo obtained its power not only as the country's most important productive center but above all as the center for linking the country's finance, information, research and development, and advanced industries to the central economies. It fills, therefore, a double role: it establishes the links with the world-economy and exercises command

Table 5.5. *Rate of urbanization in Brazil, 1950–1989*

Year	Total population	Urban population	Urban share (%)
1950	51,944,397	18,782,891	36.2
1960	70,197,370	31,533,681	44.9
1970	93,139,037	52,084,984	55.9
1980	119,002,706	80,436,409	67.6
1989	144,293,110[a]	107,239,796	74.3

[a]Excludes rural population of north region.
Source: IBGE, Demographic Censuses – 1950, 1960, 1970, and 1980; IBGE 1990b.

over the economic–financial–technological integration of the national territory as the head of its urban network.

Accelerated urbanization

Brazil became an urban country in very few decades (Table 5.5), a process which took much longer elsewhere. Some 80 million individuals have converged on the urban areas. In contrast to other countries in the southern cone which have a more stable pace of urbanization, Brazil's urbanization continues to be an extremely dynamic process. Urbanization in Brazil is the result of natural increase in the urban areas themselves; it is not, then, merely "swelling" from immigration, greater spatial mobility, or the moving frontier.

Between 1950 and 1980 the total number of cities doubled, but the most significant population growth occurred in the medium and large cities. The number of cities with more than 100,000 inhabitants increased from eleven to ninety-five over this period, representing 48.7 percent of the country's urban population in 1980. Two complementary movements characterized this urbanization: an increasing concentration in existing cities with another trend toward spatial dispersion (Davidovich and Friedrich 1988) (Figure 5.3).

In terms of *concentration*, the metropolitan regions increased their relative share of the total urban population from 25.5 percent in 1970 to 29.0 percent in 1980. Industry played a central role in the growth of these metropolises and of the urban agglomerations just short of this level. Two metropolitan regions alone, São Paulo and Rio de Janeiro with 12 million and 9 million inhabitants respectively, accounted for 75.4 percent of the occupied workforce and almost 65 percent of the country's industrial output in 1980.

São Paulo and Rio de Janeiro have been followed by two types of

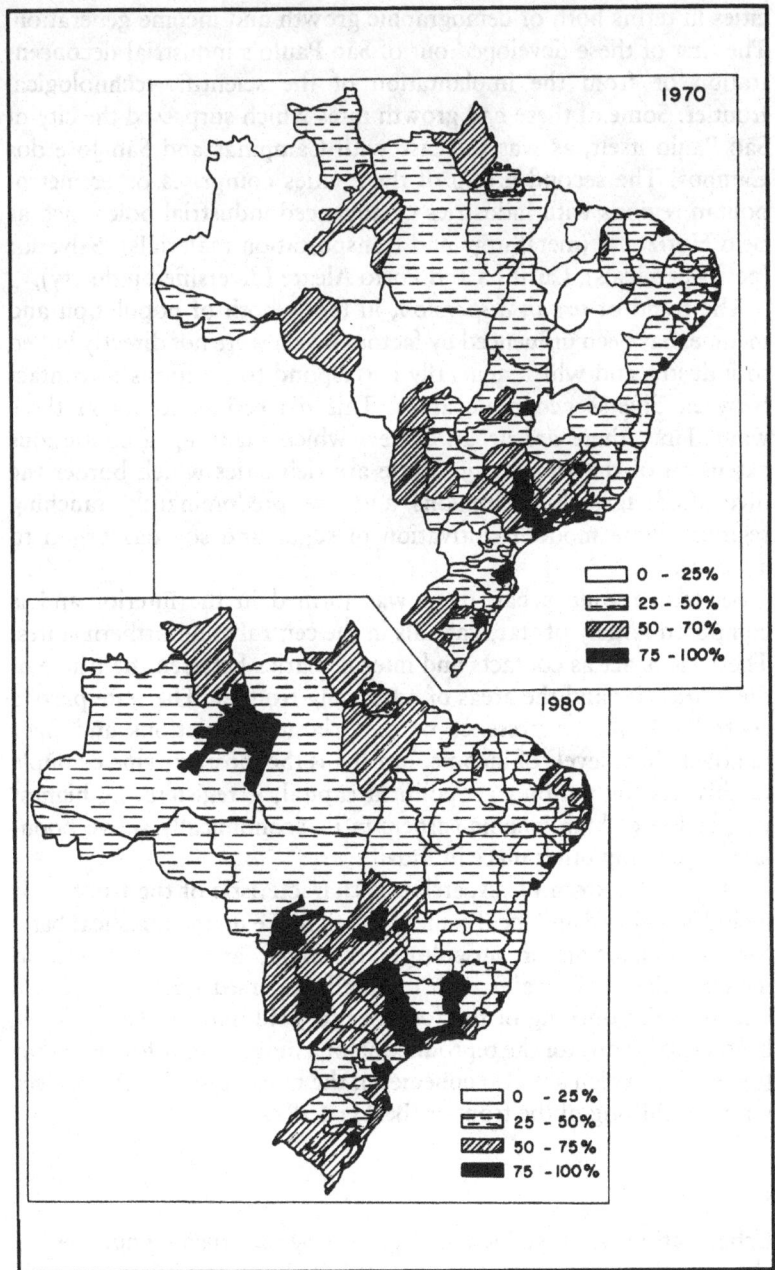

Figure 5.3 Brazilian accelerated urbanization, by meso-regions, 1970–1980.
Source: Egler 1988.

cities in terms both of demographic growth and income generation. The first of these developed out of São Paulo's industrial deconcentration or from the implantation of the scientific–technological frontier. Some of these had growth rates which surpassed the city of São Paulo itself, as was the case with Campinas and São José dos Campos. The second group of these cities comprises other metropolitan regions with industries or advanced industrial poles such as Belo Horizonte (metallurgy and transportation materials), Salvador (petrochemicals), Curitiba and Porto Alegre (diversified industry).

The trend of *urban dispersion*, in terms both of population and income, has been influenced by factors which were not directly linked to industry and which generally correspond to positions of contact between diverse economic areas. This dispersion occurs in three ways. First, there are urban centers which make up a continuous extension of the world city. These are rich cities which border the diversified agricultural regions and the predominantly ranching regions where modern cultivation of sugar and soy has begun to advance.

Second, a wide urban front was formed in the interior and is composed largely of state capitals in the central and northern states. These cities act as contacts and intermediaries between the fringe of the world city and the areas of advancing frontier. The state plays a central role in the presence of these large populations and their relatively high levels of income. Brasília is the most extreme example of this. As the nation's geopolitical capital, it registers the highest proportion of high-income earners in its economically active population than any other place in Brazil.

The third pattern of dispersion is characteristic of the frontier. It includes regional and local centers which make up the logistical basis for the expansion of agriculture, ranching, and mining. It also includes the explosive growth of small dispersed nuclei which are linked to the opening of the forest or to gold mining. These nuclei become locations for the reproduction of a mobile labor force; so that for reasons which are also ephemeral, the nuclei dislodge themselves with the shifting of the frontier (Becker 1982).

Urban poverty

Urbanization was sustained in large part by the crushing numbers of cheap and poor laborers (Santos 1979). Even so, urban work is relatively privileged since the proportion of workers in the lowest income bracket, earning up to one minimum salary, was only about

25 percent in Brazil's urban areas, much lower than the proportion of 38 percent for the country as a whole.

In spite of the multiplicity of times and spaces, a polarity of wealth/poverty between the northeast and São Paulo persists at the regional level. In the northeast, in addition to rural poverty, urbanization with induced industrialization did not result in an elevation of workers' incomes – not even in the large metropolitan centers. A northeastern pattern of urbanization exists in which low income prevails and more than 50 percent of the urban economically active population earns one minimum salary or less.

At the intra-regional and intra-urban level, the disparity further reproduces itself. The Metropolitan Region of São Paulo is much wealthier than Rio de Janeiro and great poverty is contained in the great metropolises. In the Metropolitan Region of São Paulo, the proportion of workers earning up to one minimum salary is 9.2 percent, while in Rio de Janeiro this proportion is over 14.0 percent and in Belo Horizonte it almost reaches 21 percent.

The growth of the metropolitan regions is sustained by the complementary processes of economic growth with growing poverty, spontaneous movement in the informal sector, and formal economic structures. Poverty, on the one hand, is an impediment to the expansion of the market for large companies, but on the other hand it allows for the proliferation of other factories which are less capitalized and more labor intensive. The market unifies the urban economy and the larger the city the greater the possibility of expanding informal activities. In this way the surprising expansion of industrial employment in the metropolises can be explained. Although this is employment with high turnover and low remuneration, the number of industrial jobs has increased in Brazil's metropolises in contrast to the stagnation of industrial employment in the core economies of the world-economy. In the Brazilian case, the periphery grows with industrial expansion and migration of the low-income population. The place of wealth becomes literally a place of poverty (Santos 1989).

The metropolises also become places of urban crisis, places where social deprivation of many kinds reveals itself in movements of "squatters," "invasions of the homeless," and clandestine overcrowding. They have the problems of complex administration common to the large urban agglomerations which divide into distinct local administrations, as well as the specific problems of cities in the peripheral economies. This results in a higher potential for conflicts in which people demand rights of full citizenship.

The large urban agglomerations become the principal stage for the struggles of redemocratizing society and of protecting the national industrial base against the processes of deindustrialization. The most lively expression of these struggles was the election in 1989 of the mayor of Brazil's world city, São Paulo; the victor was Luiza Erundina, a woman, northeastern migrant, and activist of the Workers Party (PT).

Complexes and networks: frames for the territory

Between 1967 and 1982, the productive structure suffered increasing transnationalization and significant changes. The authoritarian state tried to maintain high levels of investment, in distinct contrast to the policies of neighboring countries in the southern cone. This investment not only expanded the infrastructural network but also went ahead in the private sector in industrial branches which were considered strategic for the consolidation of the regime's geopolitical project.

Industrial complexes

As a result, in 1979 the industrial sector accounted for 38 percent of GNP; and manufacturing, considered the most dynamic industrial sector by World Bank criteria, accounted for 28 percent of GNP. In the last two decades, however, Brazil's industrial structure changed in important ways (Table 5.6). This transformation resulted in large part from shifts in the participation of only four industrial sectors: metallurgy and chemical products whose share increased, and textiles and food processing which declined (Penalver, Bolte, Dahlman, and Tyler 1983: 9).

In 1962, metallurgy and chemical products accounted for 20.5 percent of total industrial production, while textiles and food products reached 34.3 percent. In 1980, the situation had reversed with the former two sectors accounting for 33.8 percent and the latter for only 21.1 percent of the value of manufacturing output. The remaining manufacturing sectors essentially maintained their relative shares, with the exception of the mechanical sector, whose participation grew steadily from 2.9 percent in 1962 to 7.8 percent in 1976, declining slightly thereafter to 6.4 percent in 1980.

The effects of this dynamics, combined with the economy's own internal motion, are reflected in the territorial distribution of Brazilian industry in a contradictory fashion. On the one hand, the

Table 5.6. *Growth rates by industrial categories in Brazil, 1966–1980 (annual percentages' indices)*

Industries	1966–67	1968–73	1974–80
Consumer goods	4.8	11.9	5.0
(a) Durable	13.4	23.6	7.7
Transportation	13.1	24.0	3.3
Electrical	13.9	22.6	15.5
(b) Non-durable	3.6	9.4	4.5
Capital goods	4.5	18.1	7.1
Intermediary goods	6.8	13.9	6.8

Source: Penalver 1983.

trends of spatial concentration based on oligopolistic competition were reinforced. On the other hand, investment was also dispersed in privileged locations, such as ports, special industrial districts, and the Free Zone of Manaus (Figure 5.4).

It is important to emphasize that this movement has little to do with the emergence of regional industries which would develop relatively autonomous productive structures. On the contrary, the spatial dislocation of industrial investment which accentuated during the 1970s was a complementary process linked to the already consolidated industrial nucleus. The design of a new territorial division of labor in Brazil assumed the contours dictated by the industrial structure itself, with the accommodation of strongly integrated complexes such as the chemical and metal–mechanical ones.

Division by industrial complexes is more suited to analyzing the new territorial division of labor resulting from the insertion of Brazil as a semiperipheral nation in the world-economy (Table 5.7). First, by overcoming the artificial division between industry, agriculture, and auxiliary services, this framework permits analysis, for example, of the agro-industrial complex (AIC) which is the current form of capitalist expansion in the Brazilian countryside. Second, the concept of industrial complexes clarifies the territorial separation of management and research and development from routine production activities. While the administrative offices and centers of research and development continue to centralize, it is possible to observe the dispersion of factories across the territory depending on the required skills relative to the qualifications of the labor force.

The spatial configuration of the chemical complex in Brazil is illustrative of this process. This sector was originally concentrated in

Table 5.7. *Efficiency profile of industrial complexes, 1984 (evolution of productivity indexes)*

Complexes	Sectors			Relative size
	Ascendant	Descendant	Indefinite	
Chemical	5.1	31.7	63.2	75.9
Metal–mechanical	69.9	22.8	7.3	32.1
Agro-industrial	44.8	39.0	16.2	22.0
Textile and shoe-making	91.5	8.5	—	11.2
Paper and printing	43.3	34.4	22.1	4.5
Construction	—	60.6	39.4	4.3
Total	45.8	29.2	25.0	100.0

Note: Ascendant sector – increasing productivity from 1975 to 1984.
   Descendant sector – decreasing productivity from 1975 to 1984.
   Indefinite sector   – decreasing productivity only after the crisis of 1982.
Source: Araujo, Jr. *et al.* 1990.

the São Paulo–Rio de Janeiro axis where the large multinational corporations were established. Some of these multinationals have been in Brazil for a long time, such as Bayer and Rhodia which first established Brazilian operations in the first quarter of this century. The chemical sector has more recently expanded and diversified vigorously through massive investments, principally by state enterprises. Focusing upon the basic industries required by the chemical complex, the state decentralized the industrial plants by implanting new Petrochemical Poles – the first in Camaçari, Bahia, and the second in Triunfo, Rio Grande do Sul.

Nevertheless, the spatial distribution of the complex shows that the industry of special chemicals, intensive in technology, continues to be concentrated in the neighborhood of the world city, where firms can find the trained labor and skilled technical teams indispensable to the production of chemical substances under rigorous specifications. In the same way, the centers of management for the chemical complex remain concentrated in the Rio–São Paulo axis, whether directed by state or private enterprises. This is even so for the plants operating in the Camaçari pole, as is the case with Nordeste Quimica SA (NORQUISA) whose central headquarters is located in the financial center of Rio de Janeiro.

Similarly, the metal–mechanical complex expanded its area of activities not only around the world city but also in new locations, encouraged by the state. The axis of the automobile sector continues to be concentrated in São Paulo, with the exception of the Fiat plant

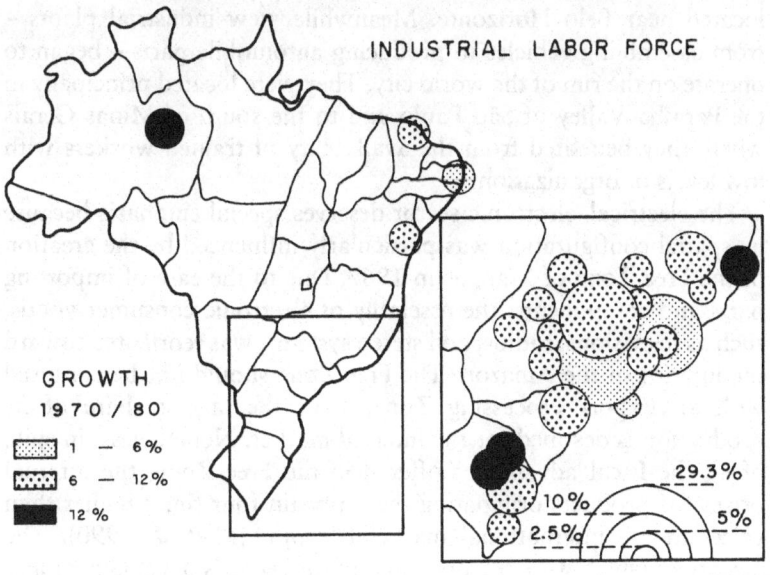

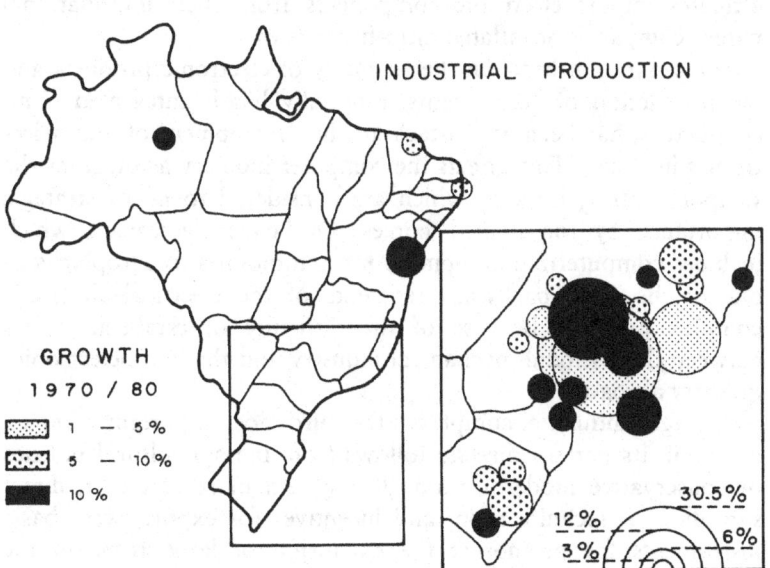

Figure 5.4 Industrial production and labor force, 1970–1980.
Source: Egler 1988.

located near Belo Horizonte. Meanwhile, new industrial plants – from assembling vehicles to producing automobile parts – began to operate on the rim of the world city. They were located principally in the Paraíba Valley of São Paulo and to the south of Minas Gerais where they benefited from the availability of trained workers with low levels of organization.

The electrical–electronic sector deserves special emphasis because its spatial configuration was particularly influenced by the creation of the Free Zone of Manaus in 1967. Due to the ease of importing parts and components, the assembly of electronic consumer goods, such as televisions, radios, and stereo systems, was reoriented toward the interior of the Amazon. The Free Zone should not be confused with an Export Processing Zone, since the largest share of its production is destined for the internal market. Nonetheless, in spite of all the fiscal advantages offered in the Free Zone, the internal prices for products originating there remain four times higher than prices in the international market (Araujo, Jr. *et al.* 1990). The incentives which were to contribute to the sector's development and expansion are instead completely transferred abroad as the Brazilian affiliates import electronic components from their multinational parent companies at inflated intra-firm prices.

The separation between the assembly of electronic products and the production of components, especially highly integrated semi-conductors, has been an obstacle to the development of microelec-tronics in Brazil. This affects the computer industry as much as the weapons sector, both of which are considered to be of strategic importance by the Armed Forces. On-board electronics, which includes computerized equipment for automobiles to aeroplanes, is one of the most backward segments of the electrical–electronic complex precisely because of the difficulty of establishing ties between the national mechanical industry and the microelectronics industry abroad.

The agro-industrial complex is the third most important complex in Brazil. Its pattern directly follows from the agricultural policies of conservative modernization. The system of rural credit, direct subsidies for technification, and incentives for export, were basic instruments for promoting the expansion of large firms in the Brazilian countryside (Table 5.8). This process converted agriculture into a necessary condition for industrial accumulation, directly linking the agro-industrial complex to the chemical and metal-mechanical ones

The structural changes are not restricted to economic or

Table 5.8. *National banking system loans to agriculture, 1973–1980 (in billions of cruzeiros)*

Period	Total	Bank of Brazil		Private banks	
		Value	%	Value	%
1972	36.6	22.9	62.4	13.7	37.6
1975	105.0	71.2	67.8	33.8	32.2
1977	212.0	154.5	72.9	57.4	27.1
1979	461.3	357.9	77.6	103.3	22.4
1980[a]	626.8	491.5	78.4	135.3	21.6

[a]January–July.
Source: Central Bank of Brazil – Economic department.

technological features alone, but incorporate social structure as well. New relations were established between rural workers, with or without land, and corporations which expanded their activities in the area (Muller 1982). Earlier ways of organizing production were reinvented under the control of agro-industrial capital. For example, *colonato* was a typical form of agricultural organization at the beginning of the century in the coffee-growing regions and has been adopted and modified in labor-intensive cultivation, such as viniculture and aviculture. Temporary and seasonal work has also become more generalized, as is the case with the *boias-frias* who inhabit the rim of small and medium cities and commute long distances to their work in the countryside.

The authoritarian manner of handling the agrarian question in Brazil was able to assure the modernization of agriculture through increasing technification while protecting large properties. The consequences of this process were inevitable: the dislocation of massive contingents of the rural population who migrated to towns and cities, functioning as a reserve labor force along with the aggravation of the historic concentration of land tenure (Tables 5.9a and 5.9b).

In the 1970s, the area cultivated by agricultural establishments expanded dramatically. This resulted from the government's fiscal and financial policies as well as the improvement of transportation – through the large road arteries which linked the areas of greatest economic development with spaces only partially integrated to production. Consequently, during the 1970s the area cultivated in Brazil increased in absolute terms by 70.7 million hectares. This was the largest increase in cultivated area since 1940 and resulted from

Table 5.9a. *Percentage of total area belonging to the 5 percent of the largest rural properties in Brazil, 1960–1980*

	1960	1970	1980
Brazil	67.9	67.0	69.3
North	90.1	64.5	68.6
Northeast	65.3	66.7	68.3
Southeast	55.2	53.0	53.9
South	56.6	56.3	57.9
Center west	64.6	67.4	65.3

Source: Hoffmann 1982.

Table 5.9b. *Land tenure in Brazil, 1985 (in % of rural establishments)*

	Less than 100 ha.		100 to 1,000 ha.		More than 1,000 ha.	
	Number	Area	Number	Area	Number	Area
Brazil	90.0	21.2	8.9	35.1	0.1	43.7
North	83.0	22.0	15.9	30.2	1.1	47.8
Northeast	94.3	28.6	5.1	39.3	0.6	32.1
Southeast	75.8	25.6	23.4	46.7	0.8	27.7
South	94.1	39.0	5.4	35.9	0.5	25.1
Center west	62.4	4.8	30.7	25.9	6.9	69.3

Source: IBGE, Preliminary Synopsis of the Agricultural Census, 1985.

strong incentives for occupation of the Amazon and the *cerrados*. The government supported this expansion with special programs for development in central and northern Brazil, created in 1975 (Mesquita and Silva 1988).

Even in the northeast, where agro-mercantile domains persist today, modernization is present in the large irrigation projects and in the revitalization of sugar cultivation, which has benefited from enormous subsidies to produce combustible alcohol with the National Program for Combustible Alcohol (PROALCOOL). The new forms of adaptation have made the regional oligarchies dependent on financing and commercialized inputs in the same way as the agricultural producers of the southeast or the south. This was achieved, however, through privileged treatment by the state apparatus – which guaranteed special lines of credit as well as captive markets for regional production. Considered as a whole, much of the conservative modernization of agriculture in Brazil can be illustrated

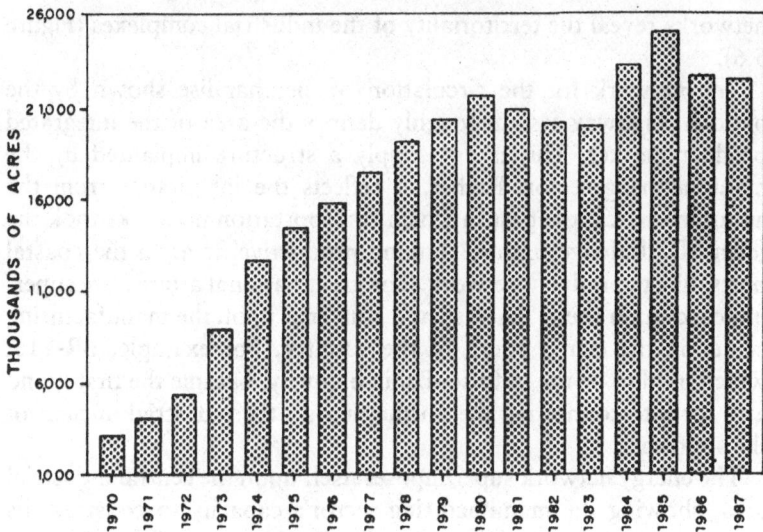

Figure 5.5 Evolution of soybean cultivation area.
Source: IBGE, Agricultural production, 1990a.

by the introduction and diffusion of soy cultivation. In 1960, some 171,000 hectares of this oil crop were planted, producing 206,000 metric tons. Twenty years later, Brazil cultivated 8.8 million hectares, harvesting 15.2 million tons (Figure 5.5). Considering the soy sector as a whole (grains, meal/cake, and oil), it came to rival coffee as the country's largest crop in value terms, accounting for 12 percent of Brazilian exports in 1980 (Homem de Melo 1983). The rapid expansion of cultivated areas, opening the ecological frontier of the *cerrados* for agriculture as well, would have been impossible without genetic improvement, massive technification, and the large scale of processing plants. Soy is a product of the new phase of Brazilian agriculture, where the agro-industrial complex plays a dominant role in molding rural space.

The national networks

The effects of the modernization process are indicated, if only superficially, by the spatial organization of the networks for the circulation of merchandise, distribution of electrical energy, and telecommunications, to the degree that they transformed past spatial structures and constructed new ones appropriate to production and administration of advanced capitalist firms. In this sense, the networks reveal the territoriality of the industrial complexes (Figure 5.6).

The network for the circulation of merchandise shown by the national highway system roughly defines the area of the integrated product market. This is not simply a structure implanted by the manufacturing sector. Rather, it reflects the inheritance from the agrarian–mercantile past in which transportation networks took the form of "drainage basins," linking productive areas to the coastal ports, like the railway network. The great national arteries are superimposed upon these "basins" and converge upon the manufacturing center in the center south of the country. For example, BR-116, which used to be the old Rio–Bahia highway, became the first grand axis for interconnecting the northeast with the industrial nucleus of the southeast.

The energy network superimposes itself upon the central industrial area, showing the manufacturing sector's capacity to construct its own territorial technical base. The distribution network for electrical energy, as a specific case, was implanted in the last thirty years by massive state investment. It is common to view energy sources as the key factor determining industrial location. In Brazil's case, however, industrialization came late, seeking to attain economies of scale from the start, and was able to benefit from the mobility of electrical energy. Brazil's industrial base, then, was built concomitantly with the energy distribution network, resulting in an extraordinary concordance between the location of industrial plants and the lines of the electrical distribution network.

The spatial results of this process can be seen when the generation and distribution system of electrical energy of the southeast is compared with its counterpart in the northeast. While the central industrial area shows a high density of power lines, forming a complex network, the northeastern system displays isolated arteries which attend to the principal urban nuclei of the region.

Finally, the national telecommunications network, expressed in the microwave system, shows that the largest urban agglomerations are interconnected in terms of rapid long-distance transfer of information. The construction of this network, begun in the 1960s and accelerated in the 1970s, shows the effects of the centralization of decision processes in the world city. This network principally attends to the demands of the financial sector which requires rapid and reliable long-distance connections in order to operate competitively.

It is important to note that the telecommunications network, from

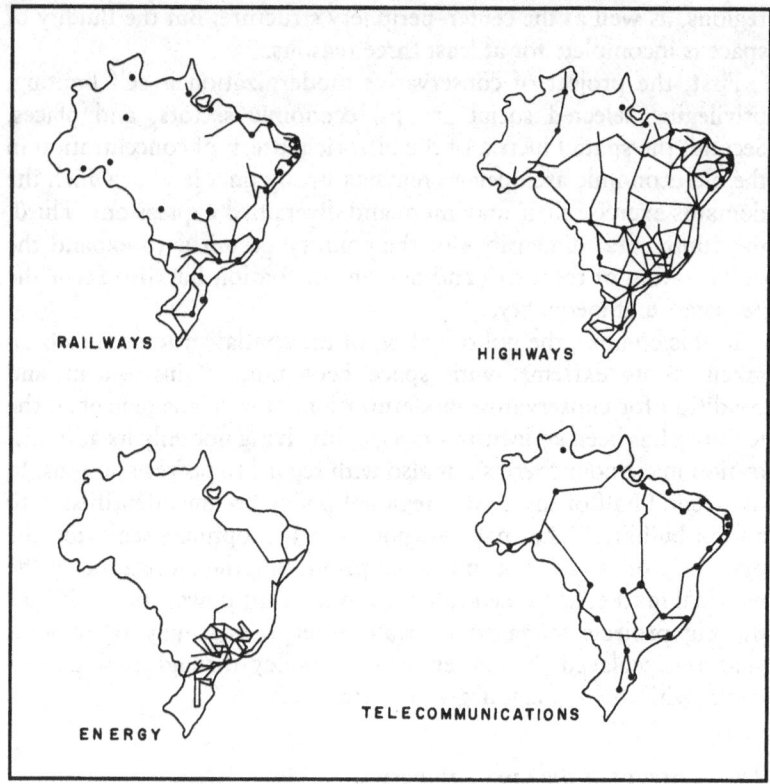

Figure 5.6 National networks.
Source: Egler 1988.

the moment of its conception, must necessarily be a national network. In short, it is the spatial expression of the most advanced form of capitalist operation, the multilocational financial firm. Only in this way is it possible to understand the rapidity with which the national long-distance communication network was developed, in only two decades, interlinking the entire national territory but leaving the majority of the population without access even to a telephone.

Transfigured space

The hybrid character of the semiperiphery is revealed in its spatial structure. The world city and the planned network tend to overwhelm the dimensions of the historic regions and the official political regions, as well as the center–periphery structure. But the fluidity of space is incomplete for at least three reasons.

First, the project of conservative modernization is self limiting, privileging selected social groups, economic sectors, and places. Second, the spatial inertia of the historic pattern of concentration in the old economic archipelago remains strong on a level at which the domains exercise their maximum and diversified expressions. Third, the continental dimensions of the country permit it to expand the mobilization of resources and human occupation but also favor the persistence of inequality.

In this context, the politicization of the spatial structure has been taken to its extreme, with space becoming an instrument and condition for conservative modernization. State management of the territory has been eminently strategic, involving not only its administration in economic terms but also with regard to power relations. In the second half of the 1960s, regional policy became identified with nation-building. The macro-region was the optimal scale for the operation of the *tripe*, as much for promoting the unification of the national market as for centralizing government power. In the 1970s, the big projects managed by state enterprises, singly or in joint ventures, replaced the earlier regional policy through new adjustments with the regional hegemonic fractions.

The spatiality of the semiperiphery

The available models of analysis for understanding a reality as complex as Brazil can be classified into two basic tendencies. The first views the system as based upon dual structures. In this conception, the traditional is opposed to the modern as a brake which hinders economic development and the diffusion of technical progress. The notion of dualism was surpassed by the concept of "structural heterogeneity," proposed originally by Anibal Pinto (1965). This concept refutes the mechanical application to Latin America of models based upon "homogeneity" of economic and social structures, typical of the central economies. Pinto's model cuts the "Gordian Knot" imposed by dualism, allowing us to understand Latin American societies not as imperfect or deformed structures, but on the contrary as societies with heterogeneity as their fundamental defining characteristic.

The question remains, how to understand the dynamics of a heterogeneous society? Or rather, since it is not "evolving" toward homogeneity, what will be its dynamic behavior? In this regard,

Wallerstein's concept of semiperiphery becomes important. As a contradictory synthesis, this concept incorporates different spaces and times in a single territory and a single moment. Political instruments, for which the state plays a central role, are key to the regulation and adjustment of these different spaces and times.

The state participated in introducing rapid changes in the contemporary world, synchronizing them with the persistence of diachronic structures whose time-flow is defined by solidly rooted patterns which tend to "slow" the clock of modernity. They are completely distinct rhythms and cadences with different velocities. They coexist in the same temporal period, requiring a complex management of the rhythms of change.

Connected to the international flows of capital, merchandise, and information, the space of flows in the semiperiphery tends to become "unglued" from the space of places based upon the permanence of territory which has been acquired historically and which strongly resists transformation. The semiperiphery is the *locus* of strong tensions which tend to bring about spatial fragmentation on various scales, generating a mosaic of modernity upon an uneven surface of misery.

Brazil is unique in exemplifying this situation, even contributing to the precision of the concept of semiperiphery. Brazil's status as a regional power was attained through a conservative modernization which produced significant transformations without breaking away from the organized hierarchical social order. The authoritarian administration of the territory was an essential instrument for producing new frontiers which broke old boundaries; securing domains which support the status quo; and consolidating the world city which is the nexus with the world-economy.

The *frontier* is not a vast extension of free lands to be exploited by people who are also supposedly free; nor is it a particular kind of periphery. Rather, it is an economic, social, and political space without a complete structure and which can potentially generate new realities. The geopolitics of the Brazilian state created not just one but many frontiers which were to open prospects for economic growth, reduce social tensions, and exercise full power over time and space.

The *domains* are consolidated areas with relatively stable political structures. They are maintained through alliances with local and regional interests who participate in the power bloc and support the conservative modernizing project. In this way, almost monopolistic forms of control over land and capital are perpetuated by all kinds of political instruments which guarantee acquired privileges and create barriers to the entrance of new competitors.

Frontiers and domains are connected by way of *the world city* which shows how Brazil is now inserted into the world-economy. The world city in the semiperiphery is, at the same time and place, the center for management and accumulation of capital on a planetary scale and the command center for a vast urban network which connects the multiplicity of spaces and times which constitute the national territory.

The emergence of the world city is explained in part by the movement of multinational capital accumulation in the world-economy. Meanwhile, the combination of this global motion with state actions creates a dynamic pattern in Brazil where the social and spatial concentration of wealth is accompanied by a selective dispersion of public and private investment by means of the planned network – which is imposed by the state, but is defined to serve the interests that make up the *tripe*.

The planned network expanded frontiers, supported domains, and strengthened the world city. At the national level, this assumed its most general expression in the persistence of the regional question in the northeast, in the configuration of an immense frontier, and in the formation of a vast urban-industrial complex from the dynamic center of the southeast. This movement expropriated and excluded significant contingents of society. It generated conflicts which are matrices for new territorialities which come to express civil society's alternative projects.

Territorial restructuring

The center–periphery structure was transfigured by conservative modernization, redefining hierarchies and power positions, restructuring functions and units of production, distribution, and management. The consolidation of the world city, of the domains, and the openings of the frontiers are expressions of this process. The new territorialities which emerge from the conflict between the planned network and life space assume specific features in each of these space–time formations (Figure 5.7).

The world city and the urban-industrial complex

The new form of Brazil's insertion into the world-economy had its greatest expression in the formation of the world city – São Paulo –

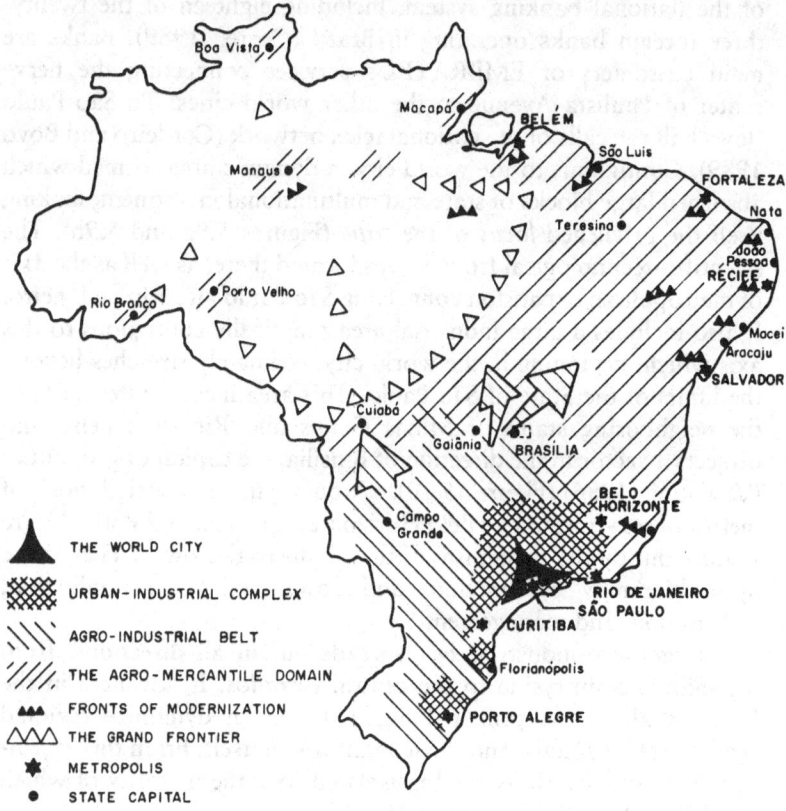

THE WORLD CITY

URBAN-INDUSTRIAL COMPLEX

AGRO-INDUSTRIAL BELT

THE AGRO-MERCANTILE DOMAIN

FRONTS OF MODERNIZATION

THE GRAND FRONTIER

METROPOLIS

STATE CAPITAL

Figure 5.7 The transfigured space.

and in the tightly linked urban-industrial structure. It emerged from concentration in and expansion of the economic nucleus of the southeast during the 1960s and 1970s. This area is the part of the country which is most integrated with the world-economy and is also the most dynamic. This is true for both its internal and external relations, promoting the territory's accelerated urbanization and generating foci of modernity. The changes in the population's territorial distribution show this process. It is characterized by the disjunction between intense urban growth, even in predominantly agricultural areas, and weak changes in population density, only significant around São Paulo and on the boundaries of the grand frontier (Figure 5.8).

Information flows are heavily concentrated in São Paulo. It is the seat of the majority of private banks which correspond to 60 percent of the national banking system, including eighteen of the twenty-three foreign banks operating in Brazil (Correa 1989). Banks are main customers of EMBRATEL's services connecting the nerve center of Paulista Avenue to the other world cities. To São Paulo flows half the calls of the national telex network (Cordeiro and Bovo 1989). Contiguous to the world city, a dynamic area formed which absorbed large blocks of state and multinational investment, making itself the privileged *locus* of the *tripe* (Figures 5.9a and 5.9b). The scientific–technological frontier was located there, as well as the axis of metropolitan expansion connecting São Paulo and Rio de Janeiro. It also includes a large industrial area practically contiguous to this axis which, beginning in the world city, ultimately stretches beyond the limits of the State of São Paulo. This area incorporates parts of the neighboring states of Minas Gerais and Rio de Janeiro and projects a vector in the direction of Brasília, the capital of geopolitics (Vesentini 1986) (Figure 5.10). Around it, a constellation of metropolises – formed by Belo Horizonte, Curitiba, and Porto Alegre – stand out by the dynamism of their industrial growth. They make up a hierarchy of functions and power linked to production, distribution, and management.

A large agro-industrial belt spreads out in all directions, from the middle countryside to the central *cerrados*. It advances in the frontiers along the principal highways which dynamize regional centers, state capitals, and the federal capital itself. From this expansion, modernity installs itself in isolated foci, the majority of which are the product of the planned network.

Pockets of poverty and conservative domains persist in the neighborhood and even within the metropolitan regions and the world city itself. On the other hand, opposition and the most important new territorialities also emerge there which cannot be disconnected from the new form of insertion into the world-economy. On the western borders of the world city where the automobile industry is located, the "new unionism" emerged autonomously from the state and with links to the international union federations. New forms of resistance to conservative modernization were generated and diffused from the world city and its immediate surroundings. Social movements with local bases represent alternative social projects in search of social justice.

The agro-mercantile domain with fronts of modernization

The agro-mercantile domain simultaneously shows the persistence of dramatic levels of rural and urban poverty along with strong

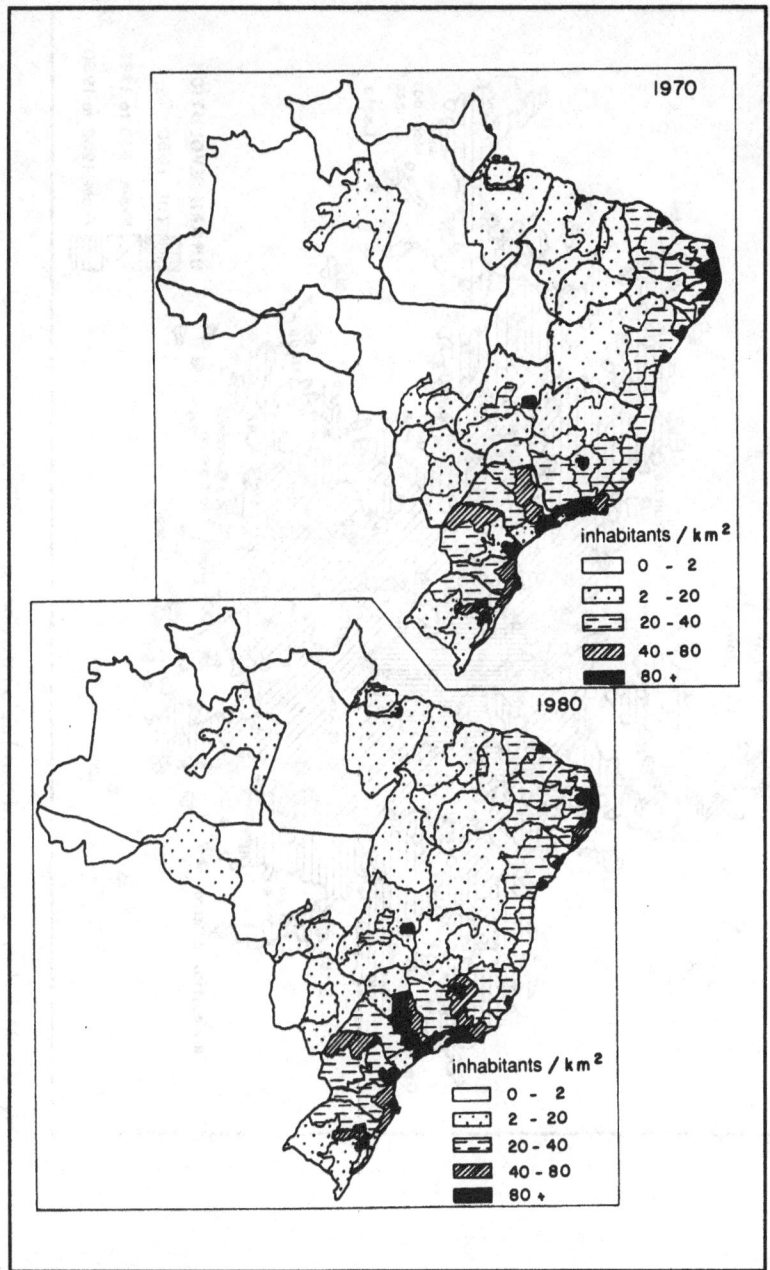

Figure 5.8 Population densities, by meso-regions, 1970–1980.
Source: Egler 1988.

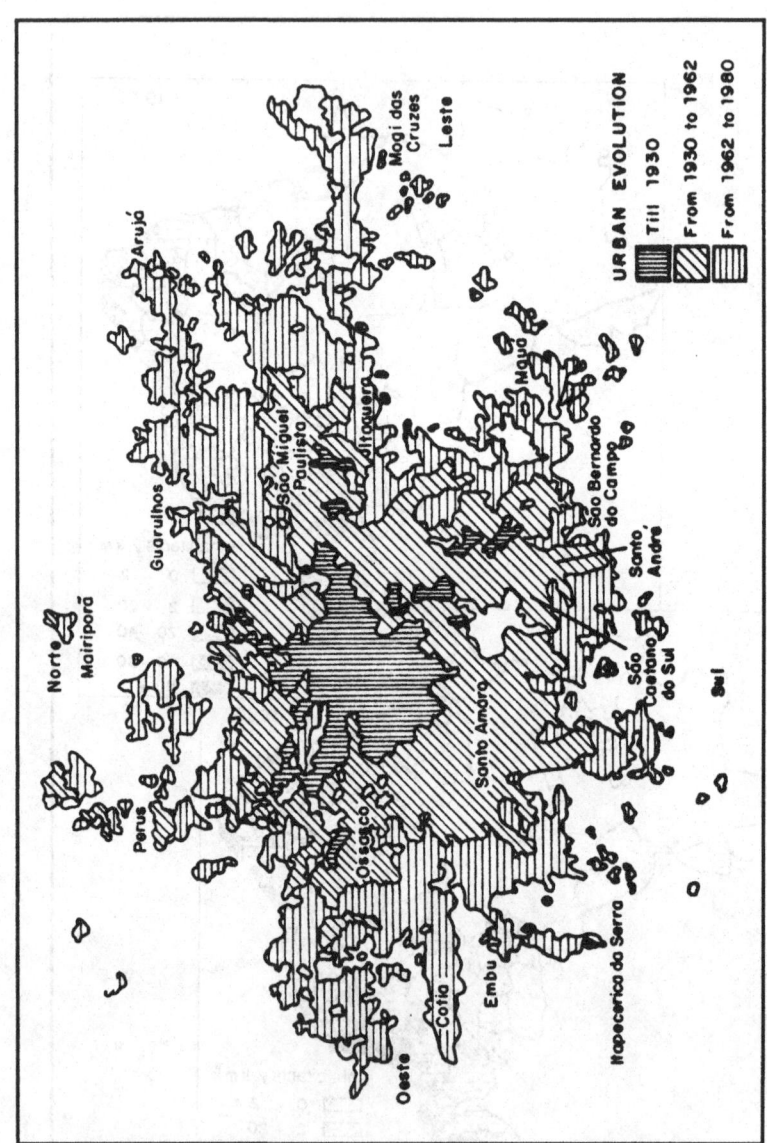

Figure 5.9a Urban expansion of São Paulo metropolis.
Source: Retrato do Brasil 1984.

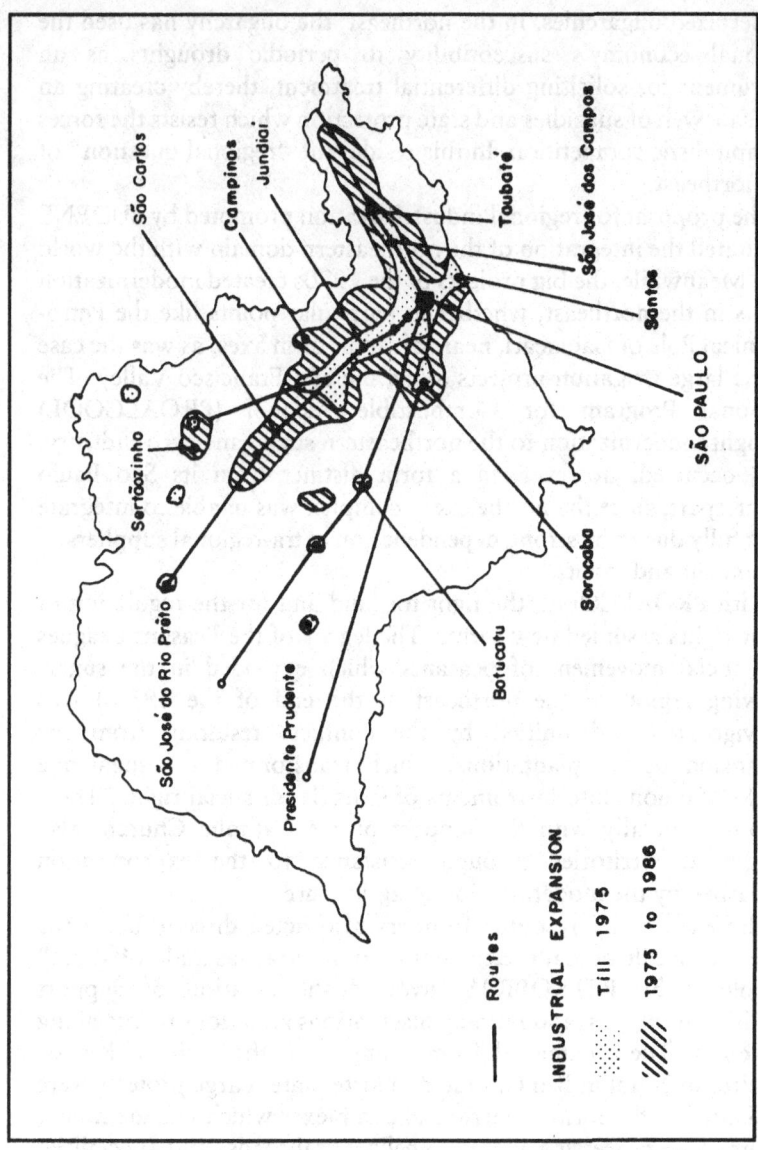

Figure 5.9b Industrial expansion in the State of São Paulo, 1975–1986.
Source: Azzoni 1989.

modernized oligarchies. In the northeast, the oligarchy has used the regional economy's susceptibility to periodic droughts as an instrument for soliciting differential treatment, thereby creating an intricate web of subsidies and state protection which resists the forces of capitalistic competition. In this resides the "regional question" of the northeast.

The proposal for regional industrialization promoted by SUDENE facilitated the integration of the northeastern domain with the world city. Meanwhile, the big projects of the 1970s created modernization fronts in the northeast, whether in particular points like the Petro-chemical Pole of Camaçari, near Salvador, or in axes, as was the case of the large irrigation projects along the São Francisco Valley. The National Program for Combustible Alcohol (PROALCOOL) brought modernization to the northeastern sugarcane agro-industry. This occurred, however, in a form distinct from its São Paulo counterpart, since the northeastern complex was unable to integrate itself fully due to its strong dependence on extra-regional suppliers of equipment and inputs.

With PROALCOOL, the fight for land and for the regulation of labor rights assumed new forms. The legacy of the Peasant Leagues (the social movement of peasants which exploded in the sugar-growing region of the northeast at the end of the 1950s) was reinvigorated and unified by the conflicts resulting from the expansion of the plantations, which transformed the sugarcane workers' unions into instruments of struggle for social rights. These unions, generally with the support of the Catholic Church, also conquered territories through resistance to the expropriation unleashed by the modernization of agriculture.

The energy and resource frontiers also acted directly upon the agro-mercantile domain. Exploitation of natural gas and "offshore" petroleum by PETROBRAS involved the creation of support facilities, terminals, and refining installations at various points along the coast. These extended from Campos, in the State of Rio de Janeiro, to Natal in Rio Grande do Norte State. Large projects were implanted in the form of territorial complexes which include mines, plantations, processing plants, pipelines, railroads, and specialized terminals in order to produce iron and ferrous metals, bauxite and alumina, caustic soda and carbonates, cellulose and paper – all primarily destined for export.

The impact of these big projects on the agro-mercantile domain was restricted. On one hand, they forced local dominant groups to reaccommodate themselves to one another, which benefited some

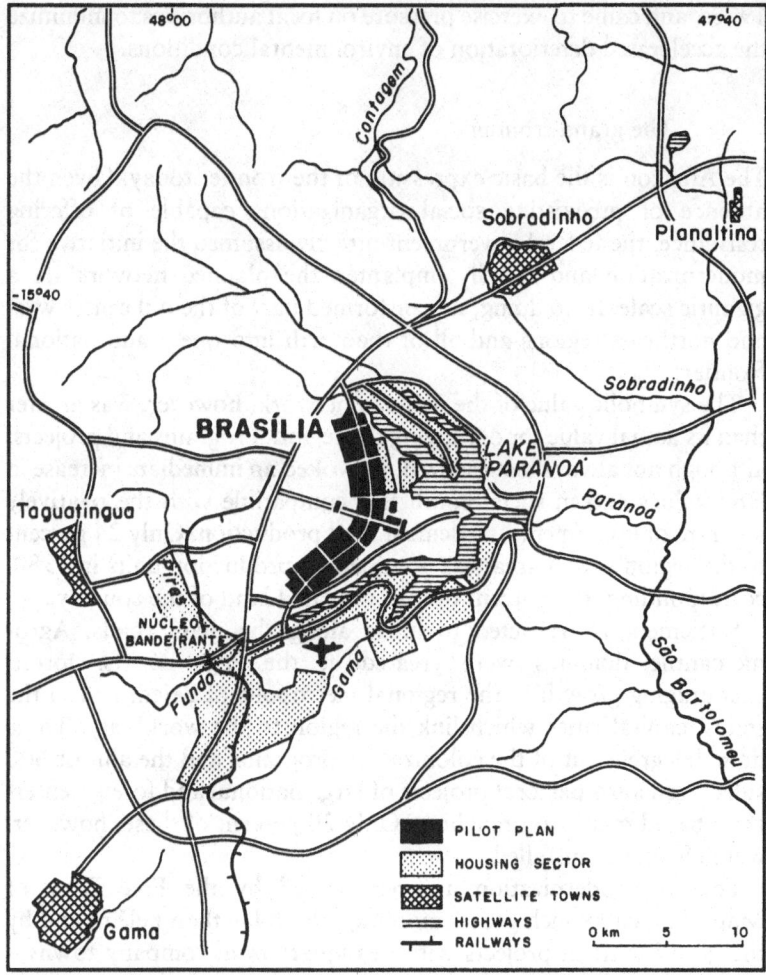

Figure 5.10 Brasília and satellite towns.
Adapted from IBGE 1966.

more than others depending on the magnitude of their resources. On the other hand, the effects were limited by the structures of production and income since the majority of these projects operated with high rates of productivity and internalized a large part of their demand for goods and services. Thus, little resulted in terms of regional development. There are, however, externalities inherent to the projects. One of these is the rapid diffusion of ecological movements which gradually assumed national extension at the end of the

1970s, and came to exercise pressure on local authorities to minimize the accelerated deterioration of environmental conditions.

The grand frontier

The Amazon is the basic expression of the frontier today. Given the absence of preexisting social organizations capable of offering resistance, the federal government directly assumed the initiative for modernization and rapidly implanted the planned network on a gigantic scale. In so doing, it transformed part of the old center west and northeast regions and all of the north into one grand national frontier.

The symbolic value of the planned network, however, was greater than its actual value for occupying the region. Programs and projects, although not always materialized, provoked an immediate increase in land values and in social conflicts incompatible with the relatively low rate of investments, settlement, and production. Only 24 percent of the region's total area was occupied by production units in 1980, corresponding to 7 percent of the cultivated land of the country.

Settlement is restricted to areas along the main roads. Agro-mercantile domains were created at the edge of the forest, encouraging growth in the regional metropolis of Belém and in the states' capital cities which link the region to the world city. These domains grew out of the colonization programs and the almost 600 subsidized agro-pastoral projects of large national and foreign enterprise based mainly on ranching. Only 20 percent of these, however, were effectively installed.

Foci of modernization are represented by the Free Zone of Manaus, a city which grows more rapidly today than Belém, and by the grand mineral projects with headquarters in company towns – nuclei of production and management – whether in isolation or in joint ventures. Due to the world recession of the early 1980s there was much less foreign investment than expected. Of the six large projects implanted under the scope of the program, only one is completely foreign – Alcoa-Billington, the largest foreign investment ever made in Brazil. Therefore the most important of these firms is the state enterprise, Companhia do Vale do Rio Doce (CVRD), which became transnational through this process, diversifying its activities and expanding its participation in the world market (Figure 5.11).

Intense social and ecological conflicts are also an outcome of the regional occupation strategy. An exponential rate of forest clearing

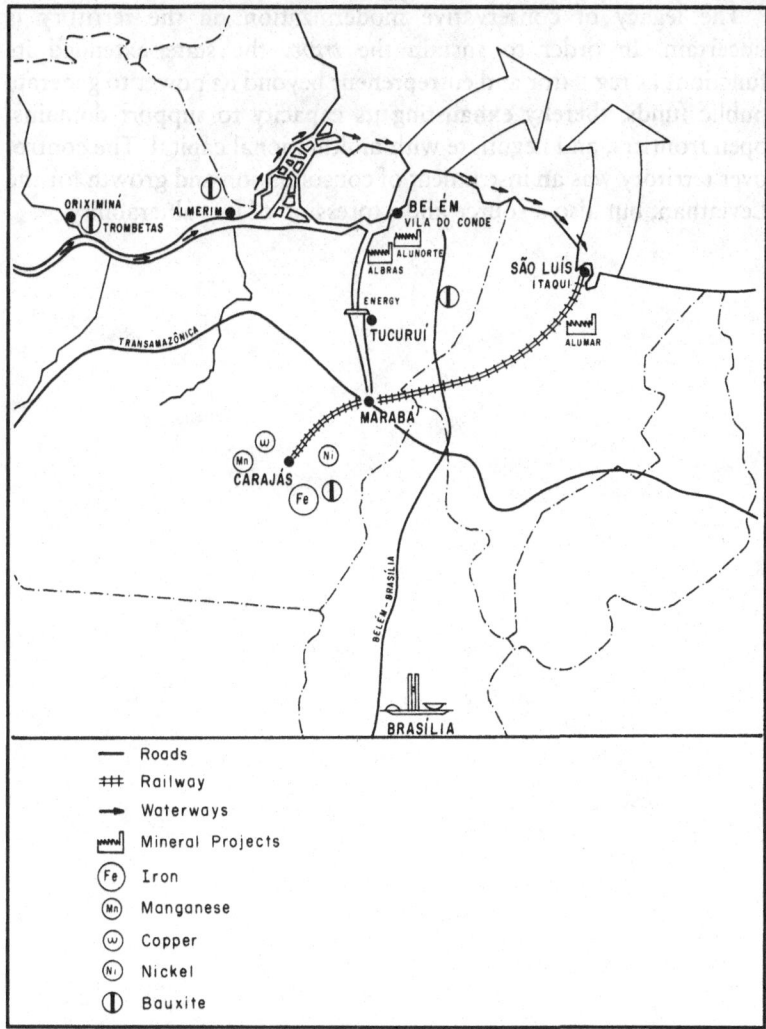

Figure 5.11 Global logistic system of the Grand Carajás Program.
Source: Becker 1990.

occurred with ranching expansion, lumbering, and mining. Estimates regarding the total cleared in the decade are conflicting and range from 12 percent (Mahar 1988), to 8.2 percent (Fearnside 1986), and 5.1 percent (National Space Research Institute – INPE). These are equivalent, respectively, to 598,921 km², 399,765 km², and 251,429 km².

The legacy of conservative modernization on the territory is uncertain. In order to sustain the *tripe*, the state extended its functions as regulator and entrepreneur beyond its power to generate public funds, thereby exhausting its capacity to support domains, open frontiers, and negotiate with multinational capital. The control over territory was an instrument of consolidation and growth for the Leviathan, but also a source and expression of its vulnerability.

6

Crisis and dilemmas of the regional power

The semiperipheral nations are intensely affected by the restructuring of the world-economy based on new technologies of production and management in forming supra-national economic spaces and in neo-liberalism as the new frame for relations between the state and the world market. At the beginning of the 1980s, the semiperiphery was directly hit by the end of a development cycle which had been sustained by external indebtedness and conducted through strong state intervention. The sharp cyclical downturn during 1981 and 1982 in the OECD countries, along with the sudden increase in international interest rates, had dramatic consequences for the semiperipheral economies.

The crisis affected the newly industrialized countries (NICs) in different ways, suggesting distinct trajectories. The Asian NICs or "tigers," smaller and more flexible, were able to recuperate more rapidly and strengthened their links with Japan, while Mexico approached the United States. The "whales," such as India, China, and Brazil – with their development based on relatively autarchic growth – remained regional powers whose dimensions are capable of refracting flows from the world-economy, although at the cost of relative isolation from the fields of influence of the world powers.

In this chapter the specificity of the Brazilian crisis is analyzed. The roots of the crisis and alternative resolutions are not predetermined by the general movement of the capitalist world-system, particularly in the regional powers, since they depend as well upon internal factors. Brazil's process of insertion into the world-economy as a semiperiphery gave it an original position. On the one hand, its links were reinforced and expanded by the massive arrival of transnational corporations, mostly North American and European. On the other hand, its national market became more autarchic relative to the world market, given the state's decisive role in regulating economic activity and forming internal income.

New forms of operation for big businesses, on a planetary scale, redefine the state's role in accumulation. These are particularly critical in Brazilian late capitalism because overcoming the crisis requires politically defining which fractions of capital will be sacrificed – implying direct negotiation among the members of the *tripe*. By taking state-building to an extreme, the legacy of military authoritarianism, in its turn, isolated itself from its social base, undermining the support necessary to negotiate in favor of nation-building.

In redefining the rules of the game, the materiality of conflict assumes clear expression in the fight for places and strategic positions in space. The geopolitical project of modernity tried to finish territorial building through imposing the planned network, extending the frontiers beyond the reach of full national control. It is then possible to reveal the interests in play within the process of economic and social restructuring through the reading of the territory. Such an evaluation will be the point of departure for analyzing the crisis and the impasses of the regional power.

The territorial dimension of the crisis

The geopolitical character of the project of modernity made the crisis show itself clearly in its territorial dimensions. It is apparent on different levels: at the local level in the fight for the right to space; at the regional level in the conflict over preserving domains; and at the national level in the dispute over control of the semiperipheral market.

The fight for space

The excessive centralization of governmental power, combined with the territorial extension of its operations, cut the communication links with the lived space, fragmenting its planned network. In spite of the multiplicity of agencies, utilized to co-opt local oligarchies, the state was unable to control resistance by the excluded population and to attend to local demands. These broke out as localized conflicts, expressed in territorially based social movements.

These movements responded to the intense expropriation and forced mobility of the labor force which tore the people from their

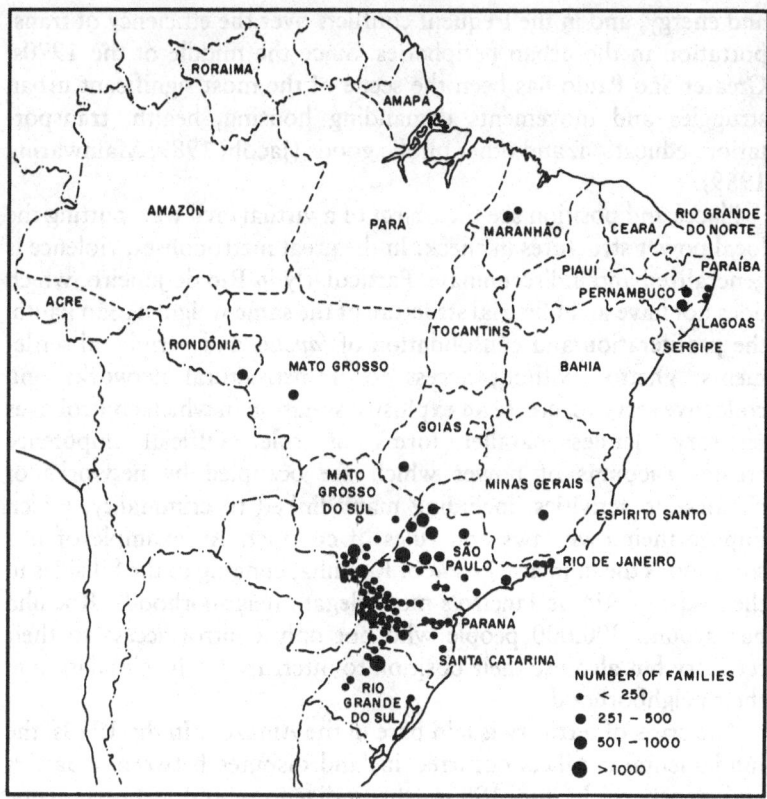

Figure 6.1 Distribution of landless camps in Brazil.
Source: Oliveira 1988.

territories, transforming the search for space into a fundamental demand among a large segment of the population. Both country and city are scenes of "invasions," which distinguish themselves from classic forms of occupation along the rim of empty lands by the velocity with which, in a single night, hundreds of destitute families create new territorialities in the heart of the great cities and valuable rural lands (Figure 6.1). The dimensions of the conflict surpass the capacities of local power to control them. In some cases, then, the conflicts erupt in armed struggle, while in others they end with the legitimation of possession.

Conquest of place is also a war of positions for access to the state's planned and selective network. The most obvious examples occur in the struggle for locations connected to the networks of circulation and energy, and in the frequent conflicts over the efficiency of transportation in the urban peripheries. Since the middle of the 1970s, Greater São Paulo has been the scene of the most significant urban struggles and movements demanding housing, health, transportation, education, and other public goods (Jacobi 1989; Mainwaring 1989).

Places and position are the object of a virtual civil war, putting the local power structures in check. In the great metropolises, violence is generalized and indiscriminate. Particularly in Rio de Janeiro, which does not have an industrial structure of the same weight as São Paulo, the proliferation and consolidation of *favelas* and peripheral settlements, ghettos without access to infrastructural networks and collective services, create an explosive situation in which control over territory defines parallel forms of rule. Official impotence creates vacuums of power which are occupied by networks of clandestine activities, including many linked to criminality, which impose their own laws and rules of conduct. An example of this situation is the immense *favela* of Rocinha, clinging to the hillsides in the midst of Rio de Janeiro's most elegant neighborhoods. Rocinha has around 300,000 people who not only control access to their territory but also use their position to interrupt traffic circulating in their neighborhood.

The crisis of territory is laid bare in the Amazon. In the 1970s, the fundamental conflicts occurred in land disputes between squatters and ranchers. In the 1980s, the indigenous and rubbertappers' communities were directly affected by the scale of the big development projects, including lumbering, which cleared vast tracts of forest. The scale of the conflicts also changed: they became disputes for territories, almost wars. Due to the magnitude of this conflict, the Union of the People of the Forest was formed in 1989 to direct the struggle for its territorialization. This union is demanding the demarcation of indigenous lands and extractive reserves – federally owned areas in which the inhabitants retain usufruct rights (Figure 6.2).

The rubbertappers of Acre use the tactic of "standoff" (*empate*) to obstruct deforestation, non-violently placing themselves in the way of the bulldozers. Their leader, Chico Mendes, was brutally assassinated, as is known worldwide, without reducing the climate of violence imposed by the ranchers. The front of *garimpeiros* continues to advance northward, threatening the lands of the Yanomami Indians, whose earth is rich in gold, tin, uranium, and precious stones. The Yanomami are also embroiled in an immense military

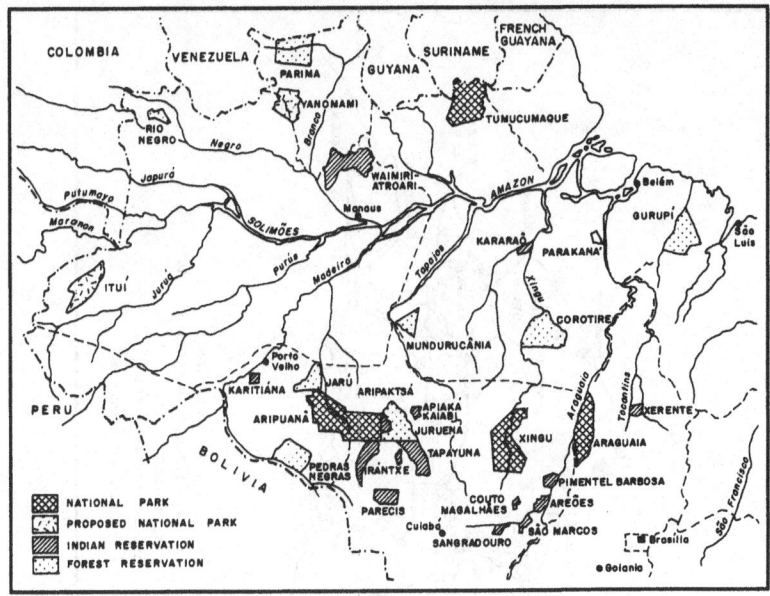

Figure 6.2 Amazon: national parks, Indian and forest reserves.
Source: Kohlhepp 1987.

project for consolidating the frontier: the Calha Norte Project, which involves 1.2 million square kilometers, some 14 percent of the national territory (Figure 6.3). The Yanomami people is subjected to all kinds of pressures, which extend from the presence of transnational religious missions to trafficking and contraband along the extensive border, military activities, and the greed of mining companies.

Conflicts also involve *garimpeiros* against firms, seeking to preserve areas for manual exploitation of minerals, as well as "people of the forest" who, dislocated from their territories, unite to demand the demarcation of indigenous lands and extractive reserves.

The Iron Project of Carajás, directed by CVRD, is a good example of this process. The power of the corporation is apparent in its control over an immense territory (2 million hectares) and the mineral reserves contained within it. CVRD maintains a citadel on its hill, surrounded by a large security perimeter. Within this territory is the Serra Pelada where more than 80,000 *garimpeiros* manually excavate the soil for gold, and have created their own citadel (Figure 6.4). The federal government was obliged to make concessions to these prospectors, extending the term for manual extraction within

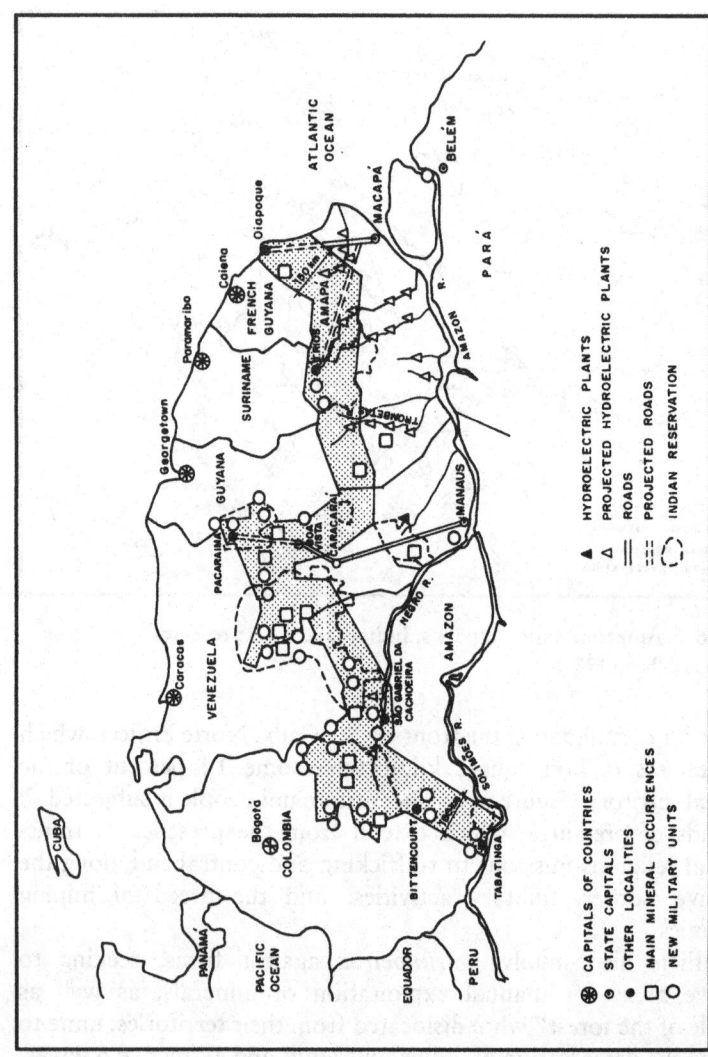

Figure 6.3 The Calha Norte Project.
Source: Becker 1990.

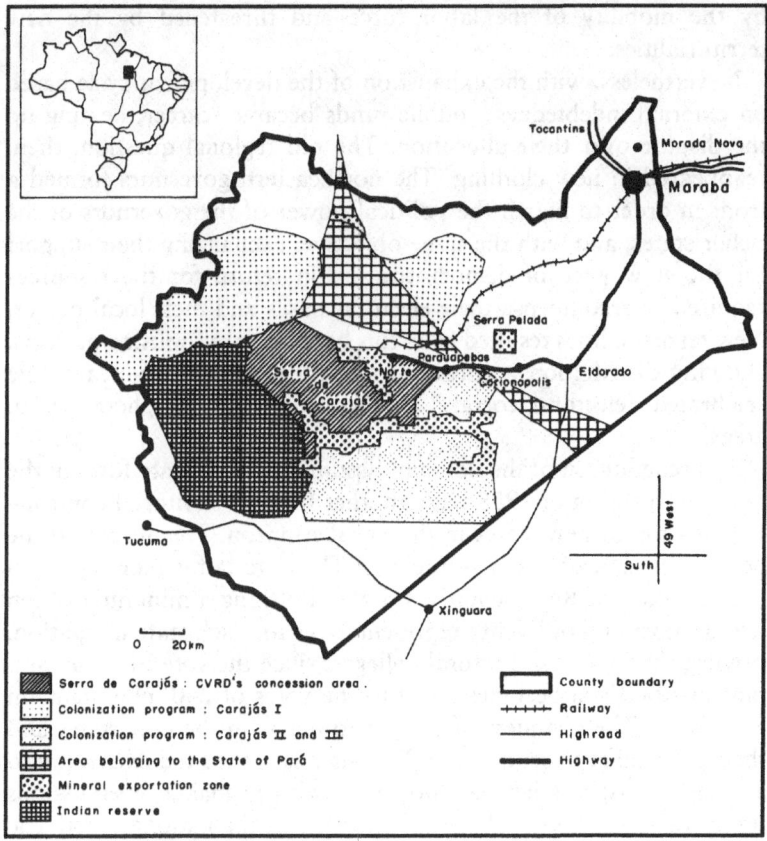

Figure 6.4 The citadel of Carajás and its security belt.
Source: Becker 1989.

the CVRD's territory in order to check a social and technological war between the company and the prospectors.

The conflict for hegemony in the domains

The slow process of authoritarian transition to civilian government (1974–85) forced a reaccommodation of the regional elites who began to seek new political mechanisms for guaranteeing their power. "Clientelism," an instrument for obtaining votes through exchanging favors and public goods, reached unprecedented levels in Brazilian history, in large part because the traditional forms of electoral loyalty – based on landownership – were profoundly shaken by the mobility of the labor force and threatened by the new territorialities.

Nevertheless, with the exhaustion of the development cycle based on external indebtedness, public funds became scarcer, heating up the dispute over their allocation. The old regional question, then, reappeared in new clothing. The northeastern governors formed a front in order to match the political power of the governors of the richer states, and with the same objective: exchanging their support for the slow pace of democratization in return for the resources required to recompense their electoral bases and their local power. New territorialities resulted from this bargain: large housing projects, allowing distribution according to political criteria, re-created the celebrated "electoral corrals" in the urban and peripheral urban areas.

The redefinition of the domains assumed an elaborate form in the new Constitution of 1988 with another form of territorial control – the creation of new states in the Legal Amazon. This increased the political weight of regional interests. The criteria for representation in the House of Representatives, by establishing a minimum of ten and a maximum of twenty representatives for each state delegation, privilege the smallest electoral colleges, since the vote of an inhabitant of Rondônia becomes equal to the votes of 240 inhabitants in São Paulo. The conquest of the governments in the new states and their political representation in Brasília has turned the Amazon into a scene of disputes between domains, with probable repercussions upon the composition of diverse interests in the national pact. The dispute for hegemony is also present in the fiscal reform proposed by the Constitution, which increased state and municipal power in administering public funds. The necessary decentralization of resources was not accompanied by a corresponding distribution of social responsibilities, which remain largely under the charge of the federal government. Predictably, given the scarcity of federal resources, the most populous states and municipalities will be penalized when confronted by the gigantic costs of their social needs, tending to aggravate the metropolitan crisis.

The territorial question

Brazil's emergence as a semiperipheral country in the world-economy altered the dimensions of its national market. Integrated industrial complexes were consolidated simultaneously with the conquest of shares of the external market because the installed capacity surpassed the level required for domestic consumption. The contradictions of the regional power were seen directly in defining its influence in the world market, whose limits depend not only upon the state's power, but also upon the competitiveness of the firms based in its territory.

Brazilian industry is extremely heterogeneous in terms of its external competitiveness, in spite of policies of incentives and tax exemptions to stimulate exports. The great beneficiaries of this policy were transnational corporations, which obtain increasing autonomy and seek to defend and expand their position by using the protection of the internal market as much as the incentives for conquering external markets. Fusions and mergers among large blocks of capital increase its power. This is the case of Autolatina, created by merging Ford with Volkswagen. In addition to holding about 50 percent of the internal market, this holding company became a large automobile exporter to Latin America, Africa, and the Middle East.

In the same way as official policy favored the conquest of positions in the external market, the location of firms was also largely induced by the state's planned network. The state pursued this as part of its regional development policy, aimed at completing the process of national economic integration. The crisis directly affected the state's capacities, not only impeding expansion, but also compromising the maintenance of its immense networks, the speed of circulation in space, and shattering of the national market. On the other hand, the large corporations which operate on a global scale aim to break the territorial limits of the nation-state in favor of places and positions, directly negotiating with local and regional fractions whose interests do not always coincide with national plans. The possibility that regional structures directly linked to the world market may resurge raises questions about the future of the regional power.

Transnational enterprises centralize management and production in the world city and its immediate vicinity. This guarantees them access to and control over the internal and external market of the semiperiphery. National firms which operate in a more competitive band of the market are more spatially dispersed (Storper 1982) and are more dependent on the directions of government policy. They may as easily consolidate their positions as be swallowed by international competition. Such risks are transparent in the proposal to implant Export Processing Zones in various parts of the territory, especially in the agro-mercantile domain. The discourse of regional development, which seeks to legitimate this project, cannot hide the desire of local interests to be directly linked to the world market, utilizing the cession of territorial parcels as an instrument for direct negotiation with transnational capital.

The biggest unknown factor in the semiperiphery resides in the future of state enterprises. Those who gained relative autonomy during the expansion of the late 1970s and consolidated their positions in the external market find themselves better situated to face the crisis. Such is the case with CVRD, which constructed its own territoriality as a government policy instrument for implementing the big projects of mineral exploration in the Amazonian frontier. CVRD created virtual enclaves which reveal the private and transnational face of a firm which directly links the region under its control to the world market (Becker 1989).

The geography of the semiperiphery expresses the complex inter-action between spaces of flows and spaces of places in the world-economy within its very territory. It also demonstrates the state's inherent difficulties in impeding the volatilization of its "national" share of the world market.

The authoritarian component of the crisis

The effects of the crisis in the 1980s directly affected the centralizing state apparatus which brought state-building ahead of nation-building to an extreme. The authoritarian military regime imposed its image of the nation upon society, eliminating the mechanisms of representation and participation which would have permitted negotiated solutions to the crisis. In its turn, the rigidity of political structures hinders new forms of decentralized and flexible administration required by the restructuring of the capitalist world-system.

The course of the semiperiphery will be defined in the political sphere and in the transitional structure of the state. Nation-building requires full conquest of citizenship, which cannot be dissociated from democratizing the state. This does not mean that the state is nothing but a cohesive force. In the process of democratization, the ambivalence is also apparent of a society that will not remain under monolithic authoritarian domination, but also lacks a tradition of political organization which is pluralistic and independent of the state (Lamounier 1988). The rhythm and the direction of political transition resulted from the clash between programmed changes, their unexpected effects, and specific social struggles.

An authoritarian transition to democracy

Economic modernization was not accompanied by political modernization. The archaic complex of practices and notions generated by the authoritarian regime and sustained by the power groups conflicted with the structural changes engendered by modernization itself and with the hierarchical rigidity of the military bureaucracy. This fragmented the state apparatus and initiated a slow and progressive process of authoritarian transition to democracy.

Liberalization in Brazil had three basic characteristics. First, it was long and gradual. It began with the "distention" which started in 1974, and became the "opening" (*abertura*) in 1979 under the control of President Figueiredo, who had been head of the National Information Service (SNI). It continued into the civilian government of the New Republic, elected indirectly in 1984, the new Constitution negotiated in 1988, and finally the 1989 direct elections for President of the Republic. Second, liberalization was the outcome of intense dialectic relations between governmental concessions and social victories, between a cautious government and a moderate opposition (Skidmore 1989). Third, in contrast to redemocratization in the southern cone, Portugal, Greece, and Spain, the Brazilian process involved the military's continuing power. The military maintained their cohesion and submitted themselves to a gradual electoral process which they themselves initiated.

It is widely agreed that the support for "distention easing" or "decompression" came from within the nucleus of the dominant system, from moderate military officers who wanted to promote a controlled liberalization capable of simultaneously attending to society's demand for change and overturning the resistance of "hard-liners." Around 1973, the formulas of state legitimation became vacuous because of the excesses of repression and increasing autonomy of the security apparatus, the increasing national and international resistance in favor of human rights, and the criticism of income inequality. The electoral path, under tutelage, was seen as an alternative with less risk. Once the process was started, the process itself produced liberalizing effects (Lamounier 1988).

One of the principal mentors of the transition was, once again, General Golbery do Couto e Silva, according to his geopolitical conception that the alternation of centralization and decentralization – systoles and diastoles – is an essential condition for the state's survival (Couto e Silva 1981). It is not by accident that among the world's forty democracies the military retain their greatest prerogatives within Brazil (Stepan 1988). Acting as a corporation, the Armed Forces control four ministries, the technical–scientific frontier, and the northern strip of the grand Amazonian frontier – the Calha Norte.

The conquests of the opposition

The viability of the electoral alternative was also the fruit of structural changes and political conquests by society. The opposition structured itself as a political force in the electoral process and ended up exhausting the military regime's pretensions to legitimate itself through elections. That is, civil society was too weak to force the beginning of liberalization, but was strong enough gradually to create informal though effective impediments to the exercise of dictatorial power. In this process, new political structures emerged which represented an advance in nation-building.

The work of articulating society was painfully woven during the military regime (Moreira Alves 1984, Sorj and Almeida 1984) escaping the immediate control of the authoritarian order. It assumed clear temporal and territorial dimensions, with the great urban centers becoming the disseminating nuclei for the opposition. The student movement was the first manifestation of resistance, leading to urban guerrilla activities at the height of the institutional violence in the years of the "miracle" growth and culminating in the guerrilla focus in the forests of the Araguaia River in the Amazon. In the darkness of this combat, human rights were suppressed and the most conservative segment of the Armed Forces, calling itself the "hard line," was strengthened.

The Catholic Church led the struggle for civil rights between 1970 and 1978. Initiated in northeastern rural conflicts, where D. Helder Camara of Recife and Olinda had a capital role, these activities were united nationally under the leadership of the Archbishopric of São Paulo, channeling the support of a part of the business class and large contingents of the middle class. Such leadership corresponded to the emergence of a new hegemony within the institution: the Church of the People, based in liberation theology. Through 80,000 Basic Christian Communities (CEBs) dispersed across the territory, the Church came to act on a national scale in the cities and the countryside, supporting the social movements that emerged within the popular classes to demand rights of access to land and citizenship (Della Cava 1989). The CEBs' roles were intense in the periphery of the big cities, with the unemployed, low-paid workers, and non-

Table 6.1. *Number of labor syndicates, 1956–1984*

Year	Urban wage earners	Professionals	Rural workers
1956	1,347	108	—
1964	1,948	120	189
1968	1,991	114	632
1974	1,949	130	1,549
1984	2,312	234	2,455

Source: *Retrato do Brazil* 1984.

unionized workers, as well as in the frontier, with expropriated peasants and indigenous communities.

Finally, after 1978, the "new unionism" emerged. This workers' movement had leaders elected within the existing union structure, but who sought to become independent of the Ministry of Labor's control. They emphasized base organization and greater militancy in strikes, even when faced with repression (Keck 1989). This movement also had a specific geographical space. It originated on the rim of São Paulo, in the ABCD region formed by the municipalities of Santo André, São Bernardo do Campo, São Caetano do Sul, and Diadema. These municipalities contain the highly concentrated automobile industry, where unprecedented strikes occurred in 1978 and 1979. From that time, a mass union structure was consolidated which spread throughout the territory, taking on national dimensions (Table 6.1).

Strengthened by the base movements, the opposition won the Amnesty Law (1979), the end of exceptional powers, and the return of party pluralism. The return of political parties redirected the social struggles toward parliamentary opposition and the base organization gradually lost its weight in civil society. This is evident in the inverse movement of the Church and the "new unionism." The hierarchy of the Church chose not to identify itself with parties, while the "new unionism," supported by intellectuals and CEB members, chose to create the Workers Party (PT) at the end of 1979 as its political expression.

The democratizing pressure only acquired expression in 1982 with the massive opposition victories in the direct elections for governors (Figure 6.5). This ended the government's monopoly on initiatives for politico-institutional legitimation. In 1984, the greatest mobilization of urban masses in Brazilian history occurred – the "Diretas Já" (Direct Elections Now) Movement. It demanded the moving forward of direct elections for the President of the Republic planned for 1988. Although this popular pressure did not attain its objective, it did further fragment the parliamentary bloc supporting the military regime, which entered negotiations with segments of the opposition. The Democratic Alliance was formed out of these negotiations and in 1984 indirectly elected the first civilian President since 1964.

Supported by a wide and heterogeneous coalition, the new government did not confront the direct resistance of the old regime, but rather the problems of its own political fragmentation. It was incapable of expressing the interests of civil society and of consolidating a new social pact for overcoming the authoritarian roots of the institutional crisis.

The national question

Contrary to classic political philosophy, the subject to be constituted and justified in Brazil is not the state but the citizen and social class (Cardoso 1979), as well as the institutions which guarantee political representation and participation. Nation-building implies confronting the majority's violent exclusion from the most elementary rights of citizenship, hindering them from becoming full members of a national community. The historic denial of land and just remuneration for labor and produced wealth, accentuated by conservative modernization, created a perverse social structure where the state's presence weighs flagrantly upon the process of forming a "regulated citizenry." The concept of citizen is not defined in the code of political values but rather in a system of occupational stratification. The citizen in Brazil is the worker whose right to work is recognized by the state (Guilherme dos Santos, 1979).

Civil society's great conquest in the struggle for full citizenship was the vote. After twenty-five years of military discretion, the right to vote was extended to the illiterate and youths of sixteen years who joined a contingent of 80 million electors, 76.9 percent of whom live in urban areas and 50.6 percent of whom earn less than one minimum wage. The electorate's low level of organization is apparent since hardly 10 percent are members of unions, only 4.4 percent are members of class associations, and 36.8 percent have no schooling (IBGE 1988b).

The behavior of this immense electorate expresses the dimensions of a society of poor masses, who project their repressed demands upon the state and from which they await concrete responses. In this context, the democratic state becomes the only organization capable

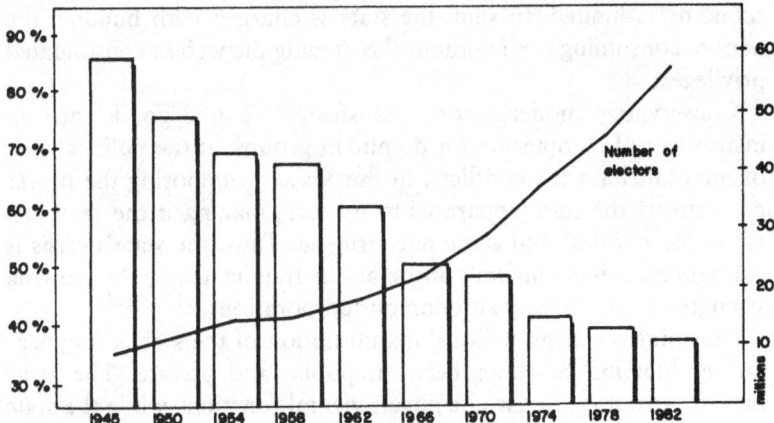

Figure 6.5 Percentage of federal deputies elected by conservative parties and electorate growth.
Adapted from *Retrato do Brasil* 1984.

of realizing that social investments and reforms on the grand scale are necessary for effectively building the nation.

The elaboration of a new Constitution and the direct election of a new President represent a new phase for political and social liberty in Brazil. In the meantime, it is marked by profound deterioration in economic relations, with the highest rates of inflation in the country's history, a brutal reduction in the levels of investment, and the steady exhaustion of the state's financial capacity.

The conquest of civil rights – the banner for wide segments of society – is included in the new Constitution, which ensures full liberty of expression and participation. Nevertheless, the constitutional assembly was conservative with respect to territorial property, making any proposal consistent with agrarian reform impossible. It also contradicted ideas of economic liberalism to the point of establishing a limit on interest rates. In this respect, the Brazilian Constitution reflects the semiperiphery's new condition, showing its ambivalence and the difficulties of forming a nation.

The crucial issue in the new Constitution is the role of the state. On the one hand, the Constitution envisions a welfare state, capable of attending to immense repressed social demands. On the other hand, it reduces the state's capacity to generate and control public funds, shared among diverse spheres of governmental power. The state is given decisive responsibility for national development, while its autonomy to create new enterprises or expand its intervention in the economy is limited. In sum, the state is charged with building the nation, consuming itself without threatening the web of consolidated privileges.

Conservative modernization transformed public goods into an instrument of co-optation for dominant groups, at the political cost of internalizing their conflicts. In this way, recomposing the power pact affects the state apparatus to its core, making it the principal arena for political and economic struggle. Thus, the state's crisis is configured, being politically and spatially fragmented by the growing strength of increasingly autonomous corporations.

Corporatism is the political manifestation of the state's fragmentation, blurring the lines between public and private. The large private corporations assume governmental functions while the state enterprises emphasize the logic of business. The military corps links itself directly to the civil sector so as to control the modern scientific and technological vector, and becomes a promoter for commercializing war materials. Faced with this situation, where political parties have little capacity for articulation, specific social struggles also assume corporatistic forms. Conflict is established between pressures to attend to particular interests and to safeguard the general interest in building the nation.

In this context, there is power but no government. Conditions are set for Bonapartism. Twenty-two parties registered for the Presidential elections in 1989, but the second round of the elections revealed a polarization between two distinct perspectives on the new role of the state. On the one side, the PT presented Luis Inacio Lula da Silva – the most highly voted-for federal deputy in Brazilian history from the State of São Paulo, metallurgical union leader, and northeastern migrant. The PT sees the state as a fundamental instrument for constructing democratic socialism in Brazil. On the other side was Fernando Collor de Melo – the young governor of the northeastern state of Alagoas and heir to a political legacy begun in the Vargas era. Without a solid party base, Collor adopted a neo-liberal discourse which was anti-statist and in favor of the millions of "shirtless" people.

Collor won with 36 million votes, against 31 million votes for Lula. In the name of economic liberalism and of a minimal state, and to combat hyperinflation, the new government's first measure was a drastic monetary reform which placed the Central Bank in a strategic position for administering the economy. It even took direct control of the balances of personal checking accounts. The economic crisis and electoral legitimation furnished the government with unlimited power to intervene in the economy and social life. This power is only comprehensible in the light of the profound political fragmentation which creates conditions for the late return of authoritarianism's populist face.

So long as the classic nation-state is not fully constituted in Brazil, the national question remains inherent in the contradictions contained in the authoritarian path of state formation. This contradiction derives from the state's determinant role in society and the economy, at the same time requiring its presence to guarantee nation-building. A long road must yet be traveled to consolidate democratic institutions and establish the full exercise of citizenship.

The late character of the crisis

Latin American economies were hit quite hard during the 1980s, considered a lost decade for the continent. Per capita income has fallen nearly 10 percent for the region as a whole. For the entire continent, average per capita income in 1985 barely exceeded the level of 1975. Unemployment surpassed 15 percent of the urban labor force in several countries; inflation has averaged 150 percent in recent years, and has exceeded 100 percent annually in half a dozen countries. Furthermore, every country carries the severe burden of external debt.

Actually, the Latin American crisis is the crisis of late capitalism, which assumes its clearest expression in Brazil. It results from the exhaustion of a particular pattern of financing; capital- and intermediate-goods industries were not accompanied by a financial structure capable of guaranteeing their expanded reproduction. Above all, it is a crisis of a state which advanced in front of the economy, not only assuming an enterprising role but also becoming the economy's principal financial agent. The state's financial crisis is apparent in the velocity of internal and external indebtedness, its inability to bank the risks inherent in the capitalist system, and its weakness when confronted with the dilemmas of new forms of integration into the world-economy.

The escalation of debt

The proximate cause of the economic crisis was the global recession in the early 1980s. The sharp cyclical downturn of 1981–82 in the United States and in the rest of the advanced industrial countries occurred jointly with the outbreak of the international debt crisis in

Table 6.2. Brazil: public-sector debt, 1982–1987 (in millions of US$)

	1982	1983	1984	1985	1986	1987
Foreign debt	48.1	62.0	63.1	69.9	81.8	90.7
Domestic debt	38.5	34.8	42.4	49.1	58.4	71.1
Total	86.6	96.8	105.5	119.0	140.3	161.8
Debt/GNP	32%	47%	50%	52%	52%	52%

Source: Central Bank of Brazil.

1982 and a decline in raw-material prices. This forced sharp reductions in imports and pushed Latin American external balances into deep deficits. Just when they were most needed, capital inflows – from both commercial banks and foreign direct investments – fell sharply.

Interest rates rose suddenly, from 1 to 2 percent annually to around 6 percent annually – levels which they had never attained before in the international financial market. These high interest rates, combined with shortened terms and high "spreads," submitted the indebted countries to draconian conditions by which to obtain new resources, many times simply to service the debt itself. In other words, the external debt gained autonomy and transformed the peripheral countries, especially in Latin America, into net exporters of capital. Between 1983 and 1988, the real transfer of resources abroad reached an annual rate of 4 percent of GNP in the Latin American countries, surpassing the effort made by Germany to pay reparations after the First World War: approximately 2.5 percent of GNP (Sampaio, Jr. 1989).

In this process, Brazil became one of the world's largest debtors in absolute terms. Its liabilities with external creditors jumped from US$40 billion in 1979 to US$112 billion in 1987, and the resources remitted abroad to service this debt represented some 5 percent of GNP. Such resources were obtained, in large part, through a rapid increase in the commercial surplus, which grew from US$1.2 billion in 1981 to US$19.1 billion in 1988 (Figure 6.6). This allowed Brazil to send abroad some US$75.6 billion in order to pay interest.

The performance of Brazilian exports, resulting from past investments, was able to diminish the crisis' effects, although it did not keep it from spreading through the national economy. The high rates of inflation, which exceeded 200 percent annually after 1982, are the most visible aspect of the profound instability which characterizes the decade's economic performance. At the root of this process is the

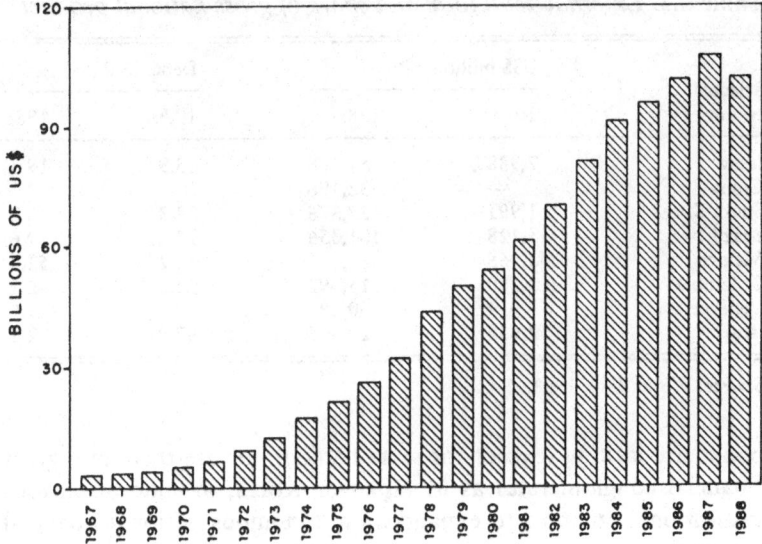

Figure 6.6 The escalation of Brazilian debt.
Basic data from the Central Bank of Brazil.

increasing indebtedness of the public sector, which reached more than 50 percent of GNP (Table 6.2). It debilitated the state's power of coordination and made any long-term economic measures impossible.

The foreign debt, and the domestic debt which was its immediate corollary, were not themselves obstacles to maintaining economic growth. Brazil had a favorable position in Latin America in terms of the relatively low ratio between foreign debt and GNP (Table 6.3). Rather, the central problem was the exhaustion of an industrialization pattern which used inflation as an instrument for solving distributional conflict – maintaining profit margins through automatically transferring any wage gains to prices. Furthermore, it counted on the public sector to socialize the risks of private investment.

The end of capitalism without risks

The dynamic of late capitalism left Brazil with a production-goods sector, but without the consolidation of a financial system capable of guaranteeing credit for long-term investment (Rangel 1987). The stock markets move a minute portion of capital assets. Investment

Table 6.3. *External debt: total and share of gross national product*

Selected countries	US$ million		Debt/GNP (%)	
	1970	1988	1970	1988
India	7,938	51,168	13.9	19.3
China	—	32,196	—	7.7
South Korea	1,991	27,376	22.3	16.2
Brazil	5,128	101,356	12.2	29.6
Mexico	5,966	88,665	16.2	52.4
Colombia	1,580	15,392	22.5	42.1
Venezuela	954	30,296	7.5	49.0
Israel	2,635	20,408	47.9	72.1

Source: World Bank 1988.

banks are lacking (except for state banks); nor are there any great financial conglomerates as in Japan or Korea, in spite of various institutional attempts to expand an autonomous national financial structure.

This effort was apparent at the start of the *Plano de Metas* with the creation of the National Bank for Economic Development (BNDE); in the "miracle" years with the National Housing Bank (BNH); and in the effort to promote financial conglomerations through the bank system during the "Forced March." Nevertheless, the great Brazilian banks grew in the context of an economy with chronic inflation and are structured to obtain large profit margins through short-term and extremely short-term operations. The efforts to consolidate a stable financial structure through capturing internal savings failed. This failure resulted in part from the ample external resources of the 1970s which linked national banks to the international financial system as intermediary institutions, recycling these funds to the internal capital markets.

Nevertheless, the process of increasing external indebtedness was only possible due to the graces of the state which also utilized foreign funds on a large scale to finance the expansion of public enterprises during the "Forced March." The formidable horizon of investments opened in 1974 was above all a political decision. The financial risks were subordinated to the logic of constructing a national power. The government gambled that the returns from future exports would be an instrument capable of settling the voluminous debts contracted with hundreds of foreign banks.

The interruption of foreign resource flows in 1982 precipitated the state's financial and economic crisis. First, the burden of external

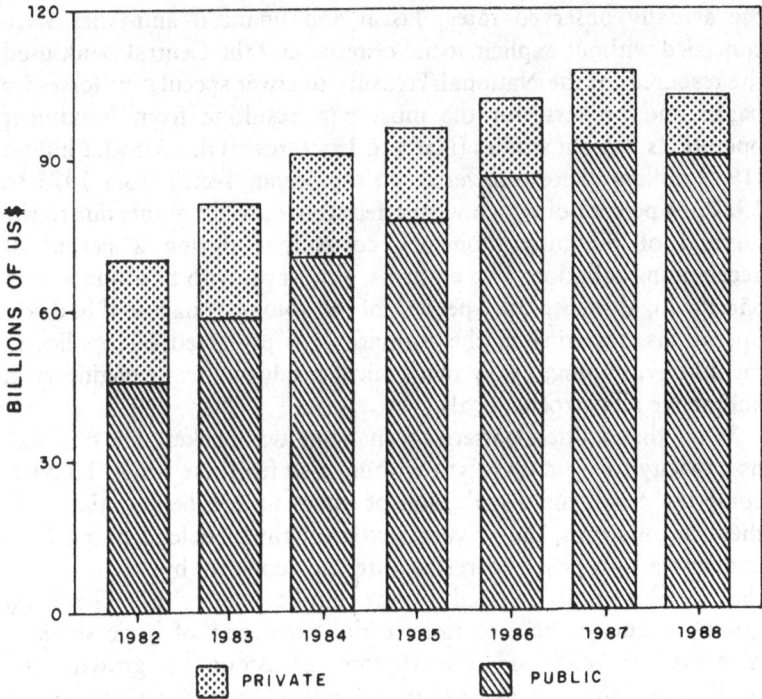

Figure 6.7 External debt: the public share, 1982–1988.
Basic data from the Central Bank of Brazil.

debt fell upon it. The Central Bank of Brazil assumed US$32 billion corresponding to public enterprise debt and US$24 billion in debt of private firms. In responding for 80 percent of the debt, it socialized the risks and politicized its negotiations with the foreign creditors (Figure 6.7). The goal of promoting growth at all costs, seeking not only to consolidate an industrial base but also its own national business class, substituted the logic of capitalist risk with the irrationality of cartels within the state. This process assumed critical dimensions with the "financial polka" (*ciranda financeira*) created by the accelerating turnover of domestic debt. It largely benefited the banking system, whose profits grew enormously as a result of speculative operations with National Treasury notes.

The transfer of public resources to the private sector was carried out through various expedients. Subsidies were granted through pre-fixing nominal rates of inflation to governmental loans, well below the actually observed rates. Fiscal and financial amnesties were conceded without explicit social criteria, and the Central Bank used the resources of the National Treasury to cover speculative losses by banks and brokers, for the most part resulting from fraudulent operations. Recent studies (Najberg 1989) reveal that US$3.2 billion (1989 dollars) were transferred to the private sector from 1975 to 1987, 74 percent of which were effectively outright grants due to the method of prefixing monetary corrections during a period of accelerating inflation. Fifteen firms, almost all with headquarters in São Paulo, absorbed 30.3 percent of the volume transferred in direct operations, resulting in the formation of protected oligopolies in metallurgy, mining, heavy mechanical products, heavy engineering, and above all petrochemicals.

When this political pattern of financing was broken, the state lost its capacity to sustain growth without risk for these firms. To overcome the crisis, the distribution of losses had to be negotiated. In these negotiations, the private sector sought to defend its position and instead proposed to restructure the economy by "privatizing" the social capital invested in the public enterprises. The great question remains how to finance the expansion of basic services, necessary to maintain the rhythm of economic growth and to improve the distribution of income – core functions of the state.

The techno(eco)logical question

In order to maintain the position attained in the international scene, Brazil must resolve the dilemma between producing at competitive costs and assuring the continual incorporation of technological innovations which demand highly risky investments – with high costs in research and development – without guarantees of immediate return. Furthermore, the world market's trends are ever more selective in terms of new products and new production processes which are part of the high technologies, principally in microelectronics, computers, precision mechanics, new materials, biotechnology, and fine chemicals.

The costs of developing these technologies are elevated not only in the separation of laboratories from production lines, but also by the necessary conditions for creating and diffusing technological innovation. Highly qualified personnel, with experience in research and development, result from investments of long maturation in which the state plays a key role. In Brazil, the case of São José dos Campos, in the scientific–technological frontier, illustrates the state's role in forming an advanced technology pole around the aerospace industry.

The crisis of late capitalism and the challenges of the regional power assume critical dimensions in this context. The financial requirements of the new pattern of insertion into the world-economy can no longer be served by socializing the risks inherent in capitalist competition. Redefining the state's role in the economy requires simultaneously guaranteeing Brazilian competitiveness in the world market for goods and services, and expanding basic social investments. For this, profound reforms are necessary to realize the potential of the semiperiphery in restructuring the world-economy.

A radical change in income distribution is fundamental to this restructuring, to break with historic privileges, expand the dimensions of the market, and overcome false dichotomies between growth "within" and "without," since the economy today is internationalized. It is also essential to interrupt the escalation of debt, to allow the consolidation of a modern financial structure capable of dynamizing not only the national banking sector but also attending to the demand for resources for autonomous development. This is particularly important for financing the production of technology and the supply of qualified services which move a significant share of international flows and are disputed vigorously by blocs of capital on a global scale.

Finally, globalization raises the ecological challenge as a question of human survival. The Amazon is a symbol of this challenge and a dilemma for the semiperiphery. The techno(eco)logical question is revealed along two contradictory expanding fronts. On one side is the energy front, which continues the geopolitical project of exploiting the frontier's resources and is led by the expansion of projects of state enterprises such as ELETRONORTE (hydroelectric energy) and CVRD (steel with charcoal). On the other side is the biotechnological front, which views nature as capital for future realization in its immense biological diversity in this greatest genetic reserve on the planet. Its preservation could mean the possibility of controlling the development of genetic engineering (Becker 1990).

The materiality of conflict involves transnational corporations and state enterprises; governmental and private financial agents; scientists and military officers; Indians, rubbertappers and *garimpeiros*, forming the strangest coalitions. It is a contradictory synthesis of national–transnational articulation and of an industrialism–

ecodevelopmentalism model dominant in the world-economy at the end of the twentieth century. Meanwhile the Amazon is not Antarctica, divided among the great world powers. It is a national patrimony and an essential element of Brazil as a nation.

7

Conclusion

The process of Brazil's insertion as a semiperipheral country in the capitalist world-economy occurs in the midst of a profound crisis which will be overcome only with a restructuring of that world-economy (Wallerstein 1983). In this transitional process, the role of the semiperipheral sector is fundamental for at least two reasons. First, the semiperiphery contains the largest share of the immense assets of the international financial system. This reduces the velocity of circulating capital and consequently reduces the introduction of innovations on a planetary scale. Second, the semiperiphery shows a sharpening of the social contradictions of historic capitalism, which contributes to stirring up the political instability of the planetary "order."

Among the semiperipheral countries, Brazil is specific. It has the largest foreign debt and the most unequal distribution of income in the world, turning its dilemmas into world challenges. The knowledge acquired about its geography not only reveals its peculiarities, but also illustrates the decisive role of the semiperiphery in the current crisis of the world capitalist system by raising two important questions. What is its pattern of insertion in the new international division of labor, modified through the formation of supra-national markets which redesign the regions of the world-economy? To what degree can its authoritarianism and its accumulated social tensions affect the direction and nature of transformation in historic capitalism?

Toward a new pattern of insertion in the world-economy

The US Vice-President Dan Quayle, after his recent visit to various Latin American countries, suggested that the opportunity is ripe for reviving continental unity given Latin America's undefined position relative to the world's new economic blocs. The "abandonment" of Latin America is notorious, a situation which would tend to aggravate with the opening of Eastern Europe to transnational capital investments. Nevertheless, it is appropriate to question what Latin America means today.

The recent cycle of growth fell unequally upon this group of countries, fragmenting them and differentiating their parts. Brazil and Mexico emerged as semiperipheral countries given their own capacities and profound links to the international financial system which also transformed them into the planet's largest debtors. Nevertheless, their prospects are quite distinct. Mexico became closer to the United States and sought to affirm its position in the Pacific Basin. Brazil, as an emerging regional power in the Atlantic, today finds its principal consumer market in the European Community, and Germany is one of its largest investors. The Europe of 1992 and a unified Germany are unknown in terms of future relations between Brazil and the central economies, placing "North–South" relations on a new footing.

The foreign debt clarifies these new relations and points out the limits of growth based on external credit, to the degree that both parts involved are responsible, and both sides end up losing. Brazil's refusal to pay the debt cannot be seen as an act of sovereignty as the Brazilian leaders would like to believe, since this merely aims at reinforcing the privileged elite which refuses to permit any reform of the social and political structure, and rejects any profound break with the international financial community. On the other hand, it is impossible to deny that Brazil is the eighth-largest industrial economy in the world, and that state expenditures sustain a class of strong industrialists who, imposing themselves on the internal and external market, are able to disquiet their competitors in the central countries.

Everything indicates that the solution is not financial but political. On the one hand, it involves the central economies' violent pressures on the "South" to adhere to the economic rules, implying their submission to the commercial constraints of the world-economy; on the other hand, it also involves a complete questioning of the false social consensus upon which the semiperiphery's governments are based in order to negotiate the debt in the name of the nation. This falsehood is transparent in all the evidence of the current incapacity to conclude a supra-national Latin American pact capable of negotiating with its creditors and of participating in the definition of new forms of its insertion into the world-economy. The root of the crisis is national, residing above all in the social and political question which is a strong obstacle to the crisis' resolution. Furthermore, it hinders the full realization of Brazil's decisive role in constructing a new continental economic space.

Toward a nation: social justice and democracy

The authoritarian path toward modernity in Brazil accentuated the historic cleavage between the minority of property owners and the mass of the dispossessed, taking credibility away from the fragile institutions which barely sustain social cohesion in the country. The indiscriminate use of public goods to benefit private groups drained the last reserves of legitimacy from the state and political activity. It leaves a legacy of profound instability which compromises the consolidation of the democratic process of constructing the nation.

To conquer misery, it is necessary to mobilize resources which only the state is capable of providing and administering. The social dimensions of planning, which passed through thirty years of conservative modernization, are the basic challenge for civil society today. In spite of the neo-liberal rhetoric, a vast social program managed by the state is essential, involving resources captured from the private sector which can no longer exempt itself from the costs of a more equitable distribution of national income.

Rescuing politics is also crucial to the conquest of citizenship. Governability depends on the affirmation of democratic institutions subject to social control and oriented toward the interests of the nation. The existence of parallel forms of power, whether they be corporatist, or clandestine, encourages the persistence of authoritarian structures which contaminate the social fabric and favor a logic of domination. The state of law is the best antidote against dictatorship in all its forms.

The Amazon incognito

As a large national/transnational frontier, the Amazon carries within it the issues pointed out above. The worldwide polemic on its ecological problem denies, at least partially, the idea of Latin America's "abandonment." Its large potential, little and inadequately utilized and still unknown, is not only a challenge to world science but also an instrument of external pressure for the adhesion to the "North" and of negotiation for Brazil against this pressure.

About one-twentieth of the earth's surface, one-fifth of the sweet water, one-third of the tropical rain forests of the globe and 3.5-thousandths of the world's population are contained in the South American Amazon and 63.4 percent of it is under Brazilian sovereignty. Despite its mineral and wood wealth, its greatest value today is the biological diversity. Such a concentration of life means on the one hand a unique ecological symbol, and on the other hand a primal source for scientific–technological development.

Therefore we have to distinguish between the myth and history. For world science, the Amazon is still an incognito. Theories about the effects of deforestation on the earth's atmospherical circulation – such as the "greenhouse effect" – are until now only hypothesis, are not confirmed, and are based upon the presumption of total destruction of the forest. Actually, despite the quick and extended clearance of the last decades, 85 percent of the forest still remains, putting forward the challenge to solve the problem of its use with preservation.

But the Amazon incognito is not restricted to science and technology. Besides the ecological problem, its geopolitical significance reflects the contradictions between the central economies in redefining their zones of influence. The proposal of converting the external debt into areas of nature conservation expresses the convergence of various sorts of interest, such as Indians, rubbertappers, national and international environmentalists, transnational companies, and governments of hegemonical powers. Large international banks, which in a recent past directly financed the "big projects," today respond not only to the environmental pressures but also to the impositions of the international agenda in times of crisis/restructuring of the capitalist world system.

The quarrel for hegemony between the world powers is unveiled in the polemic about the construction and paving of the BR-364 route which, by linking Brazil (State of Acre) to Peru, speeds up the connection with the South Pacific, where Japanese interests are stronger every day. The United States uses its influence to avoid Japanese aid for finishing the route, maintaining the traditional Amazonic open door to the Atlantic and Caribbean.

However, the difficulty in defining and negotiating a new pattern of regional development for the Amazon, considering not only the environmental dimension, but also the social problem, is also inherent in the national issue. Confronted today with a deep economic crisis, several social segments seek to consolidate positions and territories in the Amazon arena while military forces try to control the northern boundary, where the conflicts are intense. The state's fragility appears in the governmental answer to the dimension of the Amazon challenge: it is restricted to a vague proposition for an ecological–economical macro-zoning, as an instrument to regulate the fierce competition for space and to solve the ecological issue.

Brazil is an integral and inseparable part of the construction of the world-economy and its planetary dimension and its position as a semiperipheral country reveals the profound instability of historic capitalism. Finally, its dimensions as a regional power synthesize the contradictions of crisis/restructuring in the world capitalist system at the end of the twentieth century. These are lessons extracted from this book. The crucial question remains, to evaluate whether Brazil harbors the egg of the serpent or the embryo of Gaia.

Bibliography

Ablas, L. A. Q. and Fava, V. 1984. *Análise inter-regional da dinâmica espacial do desenvolvimento brasileiro*. São Paulo: FIPE/USP

Albuquerque, M. M. 1981. *Pequena história da formação social brasileira*. Rio de Janeiro: Graal

Albuquerque, M. M., Reis, A. C. F. and Delgado de Carvalho, C. 1980. *Atlas Histórico Escolar*. 7th edn. Rio de Janeiro: Ministério da Educação e Cultura

Alden, D. (ed.) 1972. *The Colonial Roots of Brazil*. Berkeley: University of California Press

Andrade, M. C. 1973. *A Terra e o Homem no Nordeste*. São Paulo: Brasiliense

Antonil, A. J. 1963. *Cultura e opulência do Brasil por suas drogas e minas*. Rio de Janeiro: FIBGE

Araujo, Jr., J. T., Haguenauer, L., and Machado, J. B. M. 1990. Proteção, competitividade e desempenho exportador da economia brasileira nos anos 1980. *Pensamento Iberoamericano*, 17, Madrid

Arruda, J. J. A. 1980. *O Brasil no Comércio Colonial*. São Paulo: Atica

Azzoni, C. R. 1989. O Novo endereço da indústria paulista. *Anais do III Encontro Nacional da ANPUR*, Aguas de S. Pedro

Bacha, E. L. 1976. *Os Mitos de uma Década – Ensaios de Economia Brasileira*. Rio de Janeiro: Paz e Terra

Bacha, E. L. and Malan, P. S. 1988. Brazil's debt: from the miracle to the fund. In *Democratizing Brazil*, ed. A. Stepan. New York: Oxford University Press

Baer, W. 1965. *Industrialization and Economic Development in Brazil*. Homewood, Ill.: Irwin

Balassa, B., Bueno, G. M., Kuczynski, P. P., and Simonsen, M. H. 1986. *Toward Renewed Economic Growth in Latin America*. Mexico City, Rio de Janeiro, Washington D.C.: El Colegio de Mexico, Fundação Getulio Vargas, Institute for International Economics

Becker, B. K. 1982. *Geopolítica da Amazônia*. Rio de Janeiro: Zahar
1984. The state crisis and the region – preliminary thoughts from a Third World perspective. In *Political Geography, Recent Advanced and Future Directions*, ed. P. Taylor and J. House. London: Croom Helm
1987. The frontier at the end of the twentieth century – eight propositions for a debate on Brazilian Amazonia. In *International Economic Restructuring and the Regional Community*, ed. H. Muegge and W. B. Stohr. Aldershot: Avebury
1988. Nation-state building in a "newly industrialized country": reflections on the Brazilian Amazonia case. In *Nationalism, Self-Determination and Political Geography*, ed. R. G. Johnston, D. Knight, and E. Kofman. London: Croom Helm
1989. Gestion du territoire et territorialité en Amazonie Bresilienne: entreprise d'Etat et "garimpeiros" à Carajás. *L'Espace Géographique*, 3, 209–17
1990. *Amazônia*. São Paulo: Atica
Becker, B. K. and Egler, C. A. G. 1989. O embrião do projeto geopolítico da modernidade no Brasil. *Texto 4*. Rio de Janeiro: LAGET
Beluzzo, L. G. M. and Coutinho, R. (eds.) 1983. *Desenvolvimento Capitalista no Brazil. Ensaios sobre a crise*. 2 vols. São Paulo: Brasiliense
BID (Banco Interamericano de Desarollo). 1982. *Socio-Economic Progress in Latin America, Annual Report*. Washington, D.C.: BID
Boxer, C. R. 1975. *The Golden Age of Brazil, 1695–1750, Growing Pains of a Colonial Society*. Berkeley: University of California Press
Braudel, F. 1979. *Civilisation materielle, économie et capitalisme, XVe–XVIIIe siècle*. Paris: Armand Colin
Bresser Pereira, L. C. 1983. Auge e declínio dos anos setenta. *Revista de Economia Política*, 3 (2)
Butterworth, D. and Chance, J. K. 1981. *Latin American Urbanization*. Cambridge: Cambridge University Press
Campanha Nacional pela Reforma Agrária. 1985. *Violência no campo*. Petropolis: Vozes/IBASE
Canabrava, A. P. 1977. A grande propriedade rural. In *História Geral da Civilização Brasileira*, vol. II: *A Epoca Colonial*, ed. S. Buarque de Holanda. 4th edn, São Paulo: Difusão Européia do Livro
Cano, W. 1977. *Raízes da Concentração Industrial em São Paulo*. São Paulo: Difusão Européia do Livro
Cardoso, F. H. 1979. On the characterization of authoritarian regimes in Latin America. In *The New Authoritarianism in Latin America*, ed. D. Collier. Princeton, N.J.: Princeton University Press
Cardoso, F. H. and Falletto, E. 1970. *Dependência e Desenvolvimento na America Latina*. Rio de Janeiro: Zahar Editores
Carvalho, J. O. 1988. *A economia política do Nordeste: secas, irrigação e desenvolvimento*. Rio de Janeiro: Campus; Brasília: ABID

Castells, M. 1985. High technology, economic restructuring and the urban-regional process in the United States. In *High Technology, Space and Society*, ed. M. Castells. Urban Affairs Annual Reviews, 28, 11–40

Castro, A. B. 1971. *7 Ensaios sobre a Economia Brasileira*. Rio de Janeiro: Forense

Castro, A. B. and Souza, F. e. P. 1985. *A economia brasileira em marcha forçada*. Rio de Janeiro: Paz e Terra

Cavagnari Filho, G. L. 1987. Autonomia militar e construção da potência. In *As Forças Armadas no Brasil*, ed. E. R. Oliveira, G. L. Cavagnari Filho, J. Q. Moraes, and R. A. Dreifuss. Rio de Janeiro: Espaço e Tempo

CECSIB. 1972. Comissão Executiva Central do Sesquicentenário da Independência do Brasil. *História do Exército Brasileiro, Perfil Militar de um Povo*. Brasília and Rio de Janeiro: Estado Maior do Exército

Comissão Pastoral da Terra. 1985. Conflitos de terra no Brasil. *Cadernos do CEAS*, 98, 16–26

Conjuntura Econômica. 1990. 44 (11)

Conrad, R. 1972. *The Destruction of Brazilian Slavery (1850–1888)*. Berkeley: University of California Press

Cordeiro, H. K. and Bovo, D. A. 1989. A modernidade do espaço brasileiro através da rede nacional da telex. *Mimeo*, UNESP: Rio Claro

Correa, R. L. 1989. Concentração bancária e os centros de gestão do território. *Revista Brasileira de Geografia, IBGE*, 51 (2), 17–32

Coutinho, L. 1987. Crise econômica e soberania nacional. In *Militares: Pensamento e Ação Politica*, ed. E. R. Oliveira. Campinas: Papirus

Couto e Silva, G. 1955. *Planejamento estratégico*. Rio de Janeiro: Biblioteca do Exército

1957. *Aspectos geopolíticos do Brasil*. Rio de Janeiro: Biblioteca do Exército

1981. *Conjuntura Política Nacional, o Poder Executivo e Geopolítica do Brasil*. Rio de Janeiro: José Olympio

Dagnino, R. 1983. *O papel do estado no desenvolvimento tecnológico e a competitividade das exportações do setor de armamentos brasileiros*. Brasília: CNPq

Davidovich, F. R. and Friedrich, O. M. B. 1988. Urbanização no Brasil. In *Brasil, uma visão geografica nos anos 1980*. Rio de Janeiro: IBGE

Dean, W. 1969. *The Industrialization of São Paulo, 1880–1945*. Austin: University of Texas Press

Delgado, G. C. 1985. *Capital Financeiro e Agricultura no Brasil*. São Paulo/Campinas: Icone/UNICAMP

Della Cava, R. 1989. The "people's church," the Vatican, and "abertura." In *Democratizing Brazil*, ed. A. Stepan. New York: Oxford University Press

ECLA (Economic Commission for Latin America). 1951. *Economic Survey of Latin America – 1949*. New York: United Nations Publications

Egler, C. A. G. 1988. Dinâmica territorial recente da indústria no Brasil:

1970–80. In *Tecnologia e Gestão do Território*, ed. B. K. Becker, C. A. G. Egler, M. P. Miranda, and R. S. Bartholo. Rio de Janeiro: UFRJ

Erber, F. S. 1985. The development of the "electronics complex" and government policies in Brazil. *World Development*, 13, 293–310

Evans, P. 1979. *Dependent Development. The Alliance of Multinational, State and Local Capital in Brazil*. Princeton: Princeton University Press

1985. State, capital and the transformation of dependence: The Brazilian computer case. *Working Paper on Comparative Development*, no. 6. Center for Comparative Study of Development, Brown University, Providence, Rhode Island

Evans, P. and Gereffi, G. 1981. Transnational corporations, dependent development and state policy in the semiperiphery: a comparison of Brazil and Mexico. *Latin America Research Review*, 16, 3

Faoro, R. 1959. *Os donos do poder*. Rio de Janeiro: Globo

Faria, V. 1988. Políticas de Governo e regulação da fecundidade: Consequências não antecipadas e efeitos perversos. *Conference on The Demography of Inequality in Contemporary Latin America*, University of Florida

Fearnside, P. 1986. Spatial concentration of deforestation in the Brazilian Amazon. *Ambio*, 15 (2)

Foweraker, J. 1981. *The Struggle for Land (A Political Economy of the Pioneer Frontier in Brazil from 1930 to the Present Day)*. Cambridge: Cambridge University Press

Franco, M. S. C. 1974. *Homens livres na ordem social escravocrata*. São Paulo: Atica

Frank, A. G. 1967. *Capitalismo y subdesarrollo en America Latina*. Buenos Aires: Signos

Furtado, C. 1959. *Formação Econômica do Brasil*. São Paulo: Cia. Editora Nacional

1970. *Economic Development of Latin America: A Survey from Colonial Times to the Cuban Revolution*. London: Cambridge University Press

Geiger, P. P. 1963. *Evolução da rede urbana brasileira*. Rio de Janeiro: Ministério da Educação e Cultura

Geiger, P. P. and Davidovich, F. 1986. The use of space by economic policies in Brazil. In A. J. Scott and M. Storper (eds.), *Production, Work and Territory. The Geographical Anatomy of Industrial Capitalism*. Boston: Edward Allen and Unwin

Germani, G. 1977. *Authoritarianism, National-populism, Fascism*. New Brunswick, N.J.: Transaction Books

Goodman, D. E., Sorj, B., and Wilkinson, J. 1985. Agroindústria, políticas públicas e estruturas sociais rurais: análises recentes sobre a agricultura brasileira. *Revista de Economia Política*, 5 (4), 31–56

Graham, D. 1972. Migração estrangeira e a questão da oferta de mão-de-obra no crescimento econômico brasileiro. *Estudos Econômicos*, 3 (1)

Graham, D. and Merick, T. 1979. *Population and Economic Development in Brazil – 1800 to the Present.* Baltimore, Md.: Johns Hopkins University Press

Guilherme dos Santos, W. 1979. *Cidadania e Justiça.* Rio de Janeiro: Campus

Hepple, L. W. 1986. Geopolitics, generals and the state in Brazil. *Colston Symposium on Geography and Politics.* Bristol: University of Bristol

Hirschman, A. O. 1979. The turn to authoritarianism in Latin America and the search for its economic determinants. In *The New Authoritarianism in Latin America,* ed. D. Collier. Princeton, N.J.: Princeton University Press

1986. The political economy of Latin American development: seven exercises in retrospection. *XIII International Congress of The Latin American Studies Association,* Boston

Hoffmann, R. 1982. Evolução da desigualdade da distribuição da posse da terra no Brasil no período 1960–80. *Reforma Agrária,* 12 (6)

Hobsbawm, E. 1977. Some reflections on the break-up of Britain. *New Left Review,* 105

Holloway, T. H. 1978. *The Brazilian Coffee Valorization of 1906.* Madison: Wisconsin University Press

Homem de Melo, F. 1983. *O Problema Alimentar no Brasil. A importância dos desequilíbrios tecnológicos.* Rio de Janeiro: Paz e Terra

IBGE (Instituto Brasileiro de Geografia e Estatística). 1967. *Panorama Regional do Brasil.* Rio de Janeiro

1988a. Perfil dos Eleitores. *Mimeo,* Rio de Janeiro

1988b. *Brasil: uma visão geográfica dos anos 80.* Rio de Janeiro

1989. *Anuário Estatístico do Brasil.* Rio de Janeiro

1990a. *Estatísticas Históricas do Brasil.* 2nd edn. Rio de Janeiro

1990b. PNAD – Síntese do Indicadores da Pesquisa Básica da PNAD de 1981 a 1989. *Mimeo,* Rio de Janeiro

IBGE/CNG, Instituto Brasileiro de Geografia e Estatística/Conselho Nacional de Geografia. 1966. *Atlas Nacional do Brasil.* Rio de Janeiro

IBGE, PNAD – Pesquisa National por Amostra de Domicílios (Domicile Sample National Research), several years

IMF, International Monetary Fund. 1985. *Directions of Trade Statistics Yearbook.* Washington, D.C.: IMF

Jacobi, P. 1989. Atores sociais e Estado. *Espaço e Debates,* 26, 10–21

Johnson, H. B. 1972. A preliminary inquiry into money, prices, and wages in Rio de Janeiro 1763–1823. In D. C. Alden (ed.), *The Colonial Roots of Brazil.* Berkeley: University of California Press

Katzman, M. T. 1977. *Cities and Frontiers in Brazil: Regional dimensions of Economic Development.* Cambridge, Mass.: Harvard University Press

Keck, M. 1989. The new unionism in the Brazilian transition. In *Democratizing Brazil,* ed. A. Stepan. New York: Oxford University Press

Klein, L. and Delgado, N. G. 1988. Recursos para a ciência: evolução e impasses. *Ciência Hoje*, 8 (48), 28–33

Kohlhepp, G. 1987. *Problemraume der Welt*, 8

Kondratieff, N. 1935. The long waves in economic life. *Review of Economic Statistics* 17, pt. 2: 105–15

Lamounier, B. 1988. Authoritarian Brazil revisited: the impact of elections upon the "abertura." In *Democratizing Brazil*, ed. A. Stepan. New York: Oxford University Press

Le Monde Diplomatique. March 1988

Lefebvre, H. 1974. *La Production de l'espace*. Paris: Anthropos

1978. *De l'Etat*. Paris: Union Générale.

Lessa, C. 1975. *15 Anos de Política Econômica*. São Paulo: Brasiliense

1979. A nação-potência como um projeto do estado e para o estado. *Cadernos de Opinião*, 15, 123–37

1990. Brasil: o futuro da questão social. *Informe CORECON*, Março

Levine, R. 1971. *Pernambuco in the Brazilian Federation 1889/1937.* Stanford: Stanford University Press

Linz, J. J. 1973. The future of an authoritarian situation or the institutionalization of an authoritarian regime: the case of Brazil. In *Authoritarian Brazil: Origins, Policies and Future*, ed. A. Stepan. London: Yale University Press

Machado, L. 1987. Intermittent Control of the Amazonian Territory. *Mimeo*, Rio de Janeiro: UFRJ

Maddison, A. 1982. *Phases of Capitalist Development*. Oxford: Oxford University Press

1985. *Two Crises: Latin America and Asia 1929–38 and 1973–83.* Paris: OCDE

Mahar, D. 1988. Government policies and deforestation in Brazil's Amazon region. *Environment Department Working Paper 7*, World Bank

Mainwaring, S. 1989. Grassroots popular movements and the struggle for democracy: Nova Iguaçu. In *Democratizing Brazil*, ed. A. Stepan. New York: Oxford University Press

Malan, P. S. 1986. Relações econômicas internacionais do Brasil (1945–1964). In *História Geral da Civilização Brasileira*, vol. XI: *O Brasil Republicano*, ed. B. Rausto. São Paulo: Difusão Européia do Livro

Manchester, K. 1973. *Proeminência inglêsa no Brasil*. São Paulo: Brasiliense

Markoff, J. and Baretta, S. R. D. 1985. Professional ideology and military activism in Brazil: critique of a thesis of Alfredo Stepan. *Comparative Politics*, 17, 175–91

Martine, G. 1989. O mito da explosão demográfica. *Ciência Hoje*, 9 (51), 28–35

Martins, L. 1985. *O Estado capitalista e burocracia no Brasil pos-64.* Rio de Janeiro: Paz e Terra

Mello, J. M. C. 1982. *O Capitalismo Tardio*. São Paulo: Brasiliense

Mendonça, S. R. 1986. *Estado e Economia no Brasil: Opções de Desenvolvimento.* Rio de Janeiro: Graal

Mesquita, O. V. and Silva, S. T. 1988. A Agricultura Brasileira: Questões e Tendências. In *Brasil: uma visão geográfica dos anos 80*, ed. DEGEO. Rio de Janeiro: IBGE

Monbeig, P. 1952. *Pionniers et planteurs de São Paulo.* Paris: Armand Colin

Moreira Alves, M. H. 1984. *Estado e oposição no Brasil.* Petropolis: Vozes

Muller, G. 1982. Agricultura e industrialização do campo no Brasil. *Revista de Economia Política*, 2 (1), 47–78

Najberg, S. 1989. Privatização dos recursos públicos – os empréstimos do Sistema BNDES ao setor privado nacional com correção monetária parcial. M.S. thesis, The Pontificia Universidade Católica (PUC), Rio de Janeiro: PUC, *Mimeo*

O'Donnell, G. 1973. *Modernization and Bureaucratic Authoritarianism: Studies in South American Politics.* Berkeley: Institute of International Studies, University of California

Oliveira, A. U. 1988. *A geografia das lutas no campo.* São Paulo: Contexto/ EDUSP

Oliveira, F. 1977. *Elegia para uma Re(li)gião.* Rio de Janeiro: Paz e Terra

Penalver, M., Bolte, E., Dahlman, C. and Tyler, W. 1983. *Política Industrial e Exportação de Manufaturados do Brasil.* Rio de Janeiro: Fundação Getulio Vargas

Pinto, A. 1965. Naturaleza e implicaciones de la heterogeneidad estructural de la America Latina. *El Trimestre Económico*, 145

Prado, Jr., C. 1945a. *Formação do Brasil Contemporâneo.* 2nd edn. São Paulo: Brasiliense

1945b. *História Econômica do Brasil.* São Paulo: Brasiliense

Rangel, I. M. 1987. *Economia Brasileira Contemporânea.* São Paulo: Bienal

Reichtul, H. P. and Coutinho, L. g. 1983. Investimento Estatal 1974–1980: Ciclo e Crise. In *Desenvolvimento Capitalista no Brasil. Ensaios sobre a crise.* Vol. II, ed. L. G. M. Beluzzo and R. Coutinho. São Paulo: Brasiliense

Reis, E. 1983. *The Nation-State as Ideology: The Brazilian Case.* Rio de Janeiro: IUPERJ

Reis, F. W. and O'Donnell, G. 1988. *A democracia no Brasil: dilemas e perspectivas.* São Paulo: Vertice

Retrato do Brasil. 1984. 5 vols. São Paulo: Política Editora

Ribeiro, D. 1977. *As Américas e a Civilização.* Rio de Janeiro: Vozes

Rodrigues, J. A. 1977. *Atlas para estudos sociais.* Rio de Janeiro: Ao Livro Tecnico

Saes, D. A. 1985. *Formação do Estado Burguês no Brasil.* Rio de Janeiro: Paz e Terra

Sampaio, Jr., P. A. 1989. Auge e declínio da estratégia cooperativa de reciclagem da dívida externa. *Novos Estudos CEBRAP*, 25, 118–35

Santos, M. 1979. *The Shared Space.* London: Methuen

1989. Por que as Metropoles explodem? *Folha de São Paulo*, February, 2

Schilling, J. 1981. *O expansionismo brasileiro*. São Paulo: Global
Schmidt, B. V. 1983. *O Estado e a política urbana no Brasil*. Porto Alegre: Universidade Federal do Rio Grande do Sul
Schwartsman, S. 1985. High technology or self-reliance? Brazil enters the computer age. In *The Computer Question in Brazil: High Technology in a Developing Society*, ed. A. Botelho and P. H. Smith. Cambridge, Mass.: Center for International Studies, MIT
Serra, J. 1979. Three mistaken theses regarding the connection between industrialization and authoritarian regimes. In *The New Authoritarianism in Latin America*, ed. D. Collier. Princeton, N.J.: Princeton University Press
1982. Ciclos e mudancas estruturais na economia brasileira do pós-guerra. In *Desenvolvimento Capitalista no Brasil. Ensaios sobre a crise*, vol. I, ed. L. G. Beluzzo and R. Coutinho. São Paulo: Brasiliense
Silva, S. 1976. *Expansão Cafeeira e Origens da Indústria no Brasil*. São Paulo: Alfa Omega
Simonsen, R. C. 1937. *História Econômica do Brasil (1500–1820)*. 2 vols. São Paulo: Cia Editora Nacional
Singer, P. 1968. *Desenvolvimento Econômico e Evolução Urbana*. São Paulo: Cia. Editora Nacional e USP
Skidmore, T. 1973. Getúlio Vargas and the Estado Novo 1937–45. What kind of regime? In *Problems in Latin American History: The Modern Period*, ed. J. Tulchin. New York: Harper and Row
1989. Brazil's slow road to the democratization: 1974–1985. In *Democratizing Brazil*, ed. A. Stepan. New York: Oxford University Press
Sorj, B. and Tavares de Almeida, M. H. 1984. *Sociedade e Política no Brasil pós-64*. São Paulo: Brasiliense
Stein, S. 1961. *Grandeza e decadência do café no vale do Paraíba*. São Paulo: Brasiliense
Stepan, A. 1973. *Authoritarian Brazil, origins, policies and fugure*. New Haven and London: Yale University Press
1988. As prerrogativas militares no regimes pos-autoritários: Brasil, Argentina, Uruguai e Espanha. In *Democratizando o Brasil*, ed. A. Stepan. Rio de Janeiro: Paz e Terra
Storper, M. 1982. The social dynamics of industrial location in Brazil: technology, labor market bargaining, income and growth. *Mimeo*. Los Angeles: UCLA
Suzigan, W. 1986. *Indústria Brasileira: Origem e Desenvolvimento*. São Paulo: Brasiliense
Tavares, M. C. 1975. *Da substituição de importações ao capitalismo financeiro*. Rio de Janeiro: Zahar
Taylor, P. J. 1985. *Political Geography. World-Economy, Nation-State and Locality*. London: Longmans
Tigre, P. 1983. *Technology and Competition in the Brazilian Computer Industry*. New York: St. Martin Press
Valverde, O. and Dias, C. V. 1967. *A rodovia Belém–Brasília. Estudo de Geografia Regional*. Rio de Janeiro: IBGE
Velho, O. 1979. *Capitalismo autoritário e campesinato*. São Paulo: Difusão Européia do Livro

Vesentini, J. W. 1986. *Brasília: a capital da geopolítica*. São Paulo: Atica

Villela, A. and Suzigan, W. 1975. *Política de governo e crescimento da economia brasileira*. Rio de Janeiro: IPEA

Viotti da Costa, E. 1966. *Da Senzala à Côlonia*. São Paulo: Difusão Européia do Livro

1977. *Da Monarquia à República: momentos decisivos*. São Paulo: Grijalbo

Virilio, P. 1977. *Vitesse et Politique*. Paris: Galilee

Waibel, L. 1958. *Capítulos de Geografia Tropical e do Brasil*. Rio de Janeiro: CNG/IBGE

Wallerstein, I. 1979. *The Capitalist World-Economy*. Cambridge: Cambridge University Press

1983. *Historical Capitalism*. London: Verso. (Translated into Portuguese as *O Capitalismo Histórico*. São Paulo: Brasiliense, 1985.)

Weffort, F. 1978. *O populismo na política brasileira*. Rio de Janeiro: Paz e Terra

World Bank. 1988. *World Development Report 1988*. New York: Oxford University Press

1990. *World Development Report 1990*. New York: Oxford University Press

Index

Most entries refer to Brazil, except where otherwise indicated. Sub-entries are in alphabetical order, except where chronological order is significant.

Printed in the United States
By Bookmasters